Into the Storm

INTO THE STORM

Donald Rawlinson

Deeds | Athens

Copyright © 2019 — Donald Rawlinson

ALL RIGHTS RESERVED — No part of this book may be reproduced in any form or by any electronic or mechanical means, including information storage and retrieval systems, without permission in writing from the authors, except by a reviewer who may quote brief passages in a review.

Published by Deeds Publishing in Athens, GA
www.deedspublishing.com

Printed in The United States of America

Cover design by Mark Babcock. Text layout by Matt King.

ISBN 978-1-947309-74-6

Books are available in quantity for promotional or premium use. For information, email info@deedspublishing.com.

First Edition, 2019

10 9 8 7 6 5 4 3 2 1

To my wife Joyce who has never given up on me. To Michelle and Heather who have never given up on their Dad. Even when Joyce was going through all the recovery trials and tribulations with me, she never threw in the towel. It is her love and God's grace that has gotten me thus far.

Thank you seems not enough.

Contents

Acknowledgements xi
Preface xiii

1. Storm of Life 1
2. Flight Training Begins 9
3. "What Am I doing?" 19
4. "I knew it would be like this!" 31
5. Are Marriage and the Army Compatible? 43
6. "Saying Goodbye?" 59
7. War or Vietnam Conflict 65
8. Medical Evacuation to Rear Echelon of Care 137
9. Try to Find a Way Out of the Storm 153
10. The realization that the Army has filed for "Divorce" 165
11. Divorce is Final—"Hero to Zero" in a Signature Line 167
12. So, what if Anything makes a Veteran Resilient 175
13. Veterans Administration 181

About the Author 189

Acknowledgements

First, I wish to thank Joe Kline for the use of his artwork for the cover of the book.

> "To best tell the soldier's story, you had to fight by his side"
> —Joe Kline.
> (Please check out his website at https://joeklineart.com/)

Second, the soldiers that I flew with were the best in the world. The resilient UH-1 "Huey" could take tremendous abuse and gunfire hits and still perform its mission.

Thirdly, the soldiers on the ground were incredibly brave and heroic. It was their willingness many times to expose themselves to danger and death just to cover us.

Fourth and not the least, all of those that came in contact with me during my lifetime and were with me in life moments. What we have shared others will never know unless they serve in combat, or life and death situations.

Fifth, to Bob Babcock who has always been a constant source of encouragement and the reason I was able to complete the book. You are a dear friend.

It is my hope that all who read the book will realize that I'm not on an endless narcissistic endeavor. It is my hope that many will read this, see some of their life, and understand that there is light at the end of the tunnel.

Preface

THIS CONTAINS INFORMATION FOUND IN MY DIARY, LETTERS TO MY WIFE, LOGbook, and memories. This will, however, be as I remember, using the Army Aviation Airmobile Concept during my time in Vietnam.

I think the whole idea of joining the Army originally was fostered by the war stories that I heard my Dad talk about to his friends. All of Dad's friends had served in the military as well as several of our family members during WWII. Francis Rawlinson had been wounded twice during the years he was overseas. He earned a Silver Star, Bronze Star, 2-Purple Hearts, and the Combat Infantryman's Badge. Therefore, he set the standard.

Into the storm is my interpretation of how I view my entrance into life in the military and the influence it has had up to and including this point in my life.

The key areas that formed my attitudes, put on hold my beliefs and challenged my very being daily for 1454 days. Or 3 years, 11 months, 24 days including the end date. Or 47 months, 24 days including the end date.

1. Storm of Life

JOINING THE ARMY — ENLISTMENT

Following the Herd mentality was the educational mantra during my high school years. It was said that if everyone does it, one can't be punished for it. How wrong a statement. Little did I realize that I would be placing a piece of my morality aside and assuming the role of "Superman" or "John Wayne".

WOMEN LOVE A MAN IN UNIFORM

I began my enlistment process in December 1964 at the recruiting office in Kalamazoo, Michigan. My recruiter was more than helpful in helping to decide on a specific area in which to serve. After going through a list of career possibilities, I asked would it be possible for me to go to flight school. He believed that I might be able to do so if I scored high enough on a battery of exams. I was overjoyed. After going to Detroit (Ft. Wayne), the battery of tests graded, and the results sealed, I returned home. After about two weeks, the good Sergeant showed up at my doorstep and came in and gave me the good news. I had scored well enough. Little did I know that he was trying to make a "quota". Next, I was sent to Ft. Wayne in Detroit for physicals, interviews, and testing. After three trips to Fort Wayne, I was accepted and scheduled to ship out after my high school graduation in June 1965. After basic training at Fort Knox, Kentucky, I was to attend Primary Flight Training at Ft. Wolters Texas.

BASIC TRAINING

Late in the afternoon on August 22, 1966, I boarded a Greyhound bus in Sturgis, Michigan enroute to Ft. Knox, Kentucky. Upon arrival much later at night in the company area, SGT Crawford met us as we disembarked from the bus. He made it quite clear that this was not going to be fun and he began barking strange words at us. (Attention, Right Face, Forward March) What do they mean? I'm tired and want to go to bed and the SGT wants to march us around in the middle of the night. What have I gotten myself into now?

What's that noise? My God, what time is it anyway? It's still dark out. Where are we going? We're going to get uniforms and such.

Ok, we are doing something different today. We start out in double time for about 20 minutes and switch to marching for about 10 minutes to this old building. We're told to remove our fatigue tops and roll up our sleeves. As I walk through the door opening I can see two medics each with pneumatic shot delivery systems shooting us in our arms simultaneously. As we walk forward, a medic on each side was giving us shots until we finished the gauntlet. Basic was fun in a strange sort of way. I enjoyed the comradery that we gained from training together. We learned many ways to take care of ourselves.

SGT Crawford had been around the Army for a long time. It was obvious from his fatigues that he had been at least an E-8 at some time. Although from the first time we met him he was an E-5. He was about 45 or more or so we thought. He would call the "Jody" and we would walk, run, duck walk, or crawl upon command. "Jody" was this mythical guy that was supposed to be taking care of our wives and girlfriends while we were away serving our country.

We were running—marching—doing push-ups or doing double-time if we were outside the building. Sgt. Crawford would have us running with our M-14 rifle, all 9.2 pounds of it, over our heads until we would begin to pass out. We would be marching down the hill, up the hill, under the hill, over the hill, around the hill, and through the hill. SGT Crawford came into the barracks one night and said to fall-out in front with full gear. He kindly informed us that we were going for a walk. I am assuming that every Army base has a hill on it affectionately referred to as "Heartbreak" or "misery" hill. We were so *overjoyed* to find that after our 10-mile walk that we had to climb the heartbreak to get back to the barracks. At 3 A.M. we arrived back at our barracks with great pride for having survived. We then prepared for our one hour of sleep.

For some of us, Hand-to-Hand Combat and pugil sticks were something of a new concept. The fighting was so totally different from the school yard fights that would happen. What really killed me was the fact that if we didn't engage with true gusto and enthusiasm, SGT Crawford would engage us and that would leave a mark or at least hurt like blazes. So, from one perspective, some of us became quite good actors. A pugil stick is a heavily padded pole-like training weapon used by military personnel for training in rifle and bayonet combat.

The grenade toss was a modified version of horseshoes and I thoroughly enjoyed participation in this event. I liked to practice until I could drop the grenade right at the base of the steel stake. We practiced with the steel bodies of inert fragmentation grenades. I loved this stuff.

I had never been hunting and I had never fired any type of firearm until I joined the Army. Ultimately, my experience with my M-14 rifle was a labor of love. I truly loved disassembling and reassembling the M-14. I would rub the wood with linseed oil until I was satisfied that it was protected from the elements. I taught myself how to dismantle the trigger group and clean the mechanism. Little did I know by learning how to do this, I would be given the task of cleaning all the M-14 trigger groups in the platoon. Now I ask you, is there such a thing as "volunteering without knowledge"? Since it was my first experience with a firearm, I wanted to know everything there was to know about it. I wasn't man enough to hold it on the target on fully automatic. It was simply too powerful a rifle. It was chambered in 7.62 mm. However, if fired one round at a time, it was a killing machine. (Note: Issued for Vietnam was the M-16 rifle—Lady Bird's gift to the infantry). Then it was off to the infirmary; he marched us for shots and more medical exams. Over the course of the ensuing weeks, we marched, ran here and there, marched some more. We marched down "misery" and up "heartbreak" hill.

I had never fired a pistol when I was growing up. I had only seen one pistol and it was in a box all taken apart. So, the first time I held such a piece of history as the 1911 .45 Automatic Colt Pistol, I was excited. It too is an extremely powerful handgun. We learned to shoot the pistol from the standing, prone, kneeling, and crouched (cowboy) positions. Note: We were issued for Vietnam the Colt .38 revolver. We considered this firearm was purposed to commit suicide.

Everyone loved the Spirit of the Bayonet "To Kill" we were practicing on the asphalt in the hot sun. Then we would attack a "dummy" on a stick. We would put the scabbards back on the bayonet and practice on each other which helped with the realism somewhat.

Everywhere SGT Crawford took us we marched and marched with "Jody". Remember, Jody was this fictitious guy back at home that would be spending time with your sweetheart or wife. Ole' Jody was a hated individual and was evil to the core and to be quite honest everyone believed that Jody was hiding when we would go home for leave.

Finally, graduation came eight weeks after we had begun the process of becoming a soldier. We marched down the street and onto the parade ground as proud US Army soldiers. Upon graduation, we returned home for two weeks leave and then to a new assignment.

INTO THE STORM

While on leave Dad took me down to a local car dealer in Vicksburg (MI), there he purchased me a Ford Fairlane. What a surprise. I was going to be able to travel in style to Ft. Wolters, Texas for Basic Flight Training. Ft. Wolters was in Mineral Wells about 80 miles west of Dallas.

Basic Training Graduation, Ft. Knox KY, 1965

DONALD RAWLINSON

Sturgis High School, Sturgis, MI Graduation Picture, 1965

Basic Training, 10 Mile Forced March 3AM, 1965

Basic Training Platoon, SGT Crofford, Drill Sargent

2. Flight Training Begins

PRIMARY FLIGHT SCHOOL TRAINING
LOCATED IN MINERAL WELLS TEXAS AT FT. WOLTERS.

"Ah, an Aviators Life"

One of my deepest desires was to be a pilot, an aviator. I drove the most direct route from Michigan and that lead me to Route 66. Route 66 was the main road from the Mid-West headed toward the West. I thought I was hot stuff. I can remember the sun, the highway, and the wind. Keeping in mind that air-conditioning back in 1965 was "2 by 55" or two windows down and 55 miles an hour. Air-conditioning on a car from Michigan was unheard of unless you were wealthy.

Arrival on November 14, 1965, to Ft. Wolters Texas.

Little did I know at the time that on this date the 7th Cavalry would engage the North Vietnamese Army in the Central Highlands of Vietnam, in the first major combat assault of the Vietnam war.

During the first month we did orientation types of activities, character building, push-ups, sit-ups, marching, running and more marching, getting our uniforms, helmets, books, and training supplies, barrack assignment, and rooms, special items for the wall locker inspections. Inspection after inspection, the TAC officers loved inspections. If they weren't inspecting the rooms they were inspecting "you". They would walk around with rulers in their hands and stop you, call you to attention and measure the distance from the ground to the bottom of your bloused fatigues. If the distance was not 10 inches precisely, you received demerits and could not leave the post or the area on the weekends.

We received more inoculations in Texas. It was required that we run everywhere and there were no exceptions. If we got caught not running, we did pushups. Cleaning our room was a precise operation. Making sure all our clothes were in a specific order on hangers. When in formation, we were "NOT" to look at any helicopters that were flying. We were to keep our eyes straight ahead and not move (discipline). If we failed to keep our eyes straight ahead, the "Tac" Officer would tell us to do one of two things. One, we could lie on our back and assume the dying cockroach position and cry out, "I see a helicopter." Two, we would be told to go to the "box" and march around the box until we saw a helicopter. At which point, we were to stop, come to attention, turn and face the approaching helicopter and raise one's arm. We would then point to the helicopter and cry out, "I see a helicopter" until it flew out of sight. One can only imagine how many times one could stop before two hours were up. The additional duties they would put us on ranged from mowing the grass with scissors to cleaning the underside of the commode with a toothbrush for hours.

The first 30 days was focused on concentrating on the essentials such as billeting, base familiarity, and drawing equipment from the supply. Yep and oh-by-the-way we needed more shots here. More focus on the quality of our dental work since it affects one's ability to fly. Ok, how can a cavity affect one's flying ability? As altitude changes, pressure changes occur thereby producing pain in one's mouth. That duress is simply dangerous while being under the stress of initial flight training where someone doesn't care if you succeed or fail. A short list of some of the classes we were required to master.

- Federal Aviation Regulations, FAR's
- Aerodynamics
- Aeronautics
- Navigation
- Meteorology
- Weather Briefings
- Weather Decision Making
- Weather Symbology
- Radios Operations (Communications, Navigation)
- Communication Procedures
- Morse Code
- Emergency Procedures (Survivable and Non-survivable)
- Flight Training

After the first 30 days, we were assigned the aircraft type in which we would be training for the next four months. My training would be in the Fairchild-Hiller OH-23D, G models. The "G" model had more power. My very first flight in anything that went higher than a step ladder was the orientation flight at Ft. Wolters. The pilot added RPM and collective pitch and the aircraft began to climb. I am sure that my eyes were as large as saucers. It was a beautiful experience seeing the ground pull away further and further as the aircraft climbed to an indicated altitude of 2,000 feet. The instructor never ceased speaking to me as he flew the aircraft. He was outlining his every move and how the aircraft was responding to his actions. After the aircraft leveled off at 2,000 feet, he said to take the controls and just try and maintain the 2,000-foot altitude. Now it should be said that this helicopter had a cyclic, collective, two pedals, motorcycle type throttle grip and about 15 instruments that you were dealing with all the time you were flying and watching where you were going. I was given about 45 minutes of flight time which was enough to convince me that this is what I wanted to do in the Army. Under "dual" flying, we were flying with an instructor pilot.

Now the next day, the blasted helicopter didn't want to do anything that you wanted it to do. It was if the helicopter was demon possessed. You had to plan ahead into next week just to get it to hover. My first experience with hovering the blasted thing was in a 40-acre field. (I'm not kidding.) My instructor pulled it up to a 20-foot hover and said just try and keep it in the field and don't kill us. A piece of cake right!?? Absolutely NOT! It went where it wanted, and I was finally relieved when he took control of the aircraft. I thought he was going to let me kill both of us. He had nerves of steel. I just had a case of nerves. The true difference is that he knew the limitations of the aircraft and I had no concept of my limitations or the aircraft.

On one day, my instructor looked at me and said, "Well, now you've finally done it." With that, he told me to go around the pattern three times and perform three solo auto-rotations. I was in shock, moments passed, then that feeling came over me, you know the one, "Ah sir, don't you think I need to work on the auto-rotations a little longer?" He muttered something about not embarrassing him. Keep in mind the washout rate was 40%. So, I figured I had made it to the point that I wouldn't wash out, just kill myself.

I managed to make all three of the auto-rotations without crashing. Well, at least I didn't damage the "Raven". They were rough and too fast but survivable. Actually, I thought they were all crashes, but they were good crashes. I managed to walk away from all three of them.

The trip by bus back to the post was interrupted by a stop by a frozen water hole. The guys all broke a hole in the ice by throwing me into the water hole. (Tradition for those that solo, however, it should be noted that some got thrown into the Officers Club swimming pool.) Nope, I got thrown in with all the cow paddies. The pool was for the civilized soldiers. You know—the commissioned officers and their families. We used to make fun of the fact that Warrants weren't "real" officers we were just experts in what we were trained to do.

Since I soloed, this meant that I could wear my hat correctly. I could now turn it around. I did not have to wear it backward any longer. Hooray! I've graduated from a scum sucking maggot to a lower life form not previously identified. It was a tradition that until a student soloed, they had to wear their hat backwards. Its purpose was to help motivate us to work harder to solo.

In 1965 and beginning of 1966, our stage fields (miniature airports) where we practiced were numbered and not named. Stage field 4 is etched into my mind since we had an accident that occurred there while I was solo. A dual (instructor and student) aircraft landed on top of one of the Solo (student only) aircraft. One might ask how that could happen. The solo aircraft didn't have communication radios in them. The radios were needed overseas and as such weren't available. Even though the Tower could tell that the solo aircraft was in danger, the Tower could not radio the student to make him aware of the impending situation. The instructor pilot in the dual aircraft was concentrating on the instruction being given to the student doing the approach. The solo aircraft was in a blind spot and could not be viewed by the instructor in the dual aircraft. When the dual aircraft impacted the solo aircraft, he hovered for a moment before the solo aircraft sliced into the fuel tank of the dual aircraft and both struck the ground in flames. The solo student was out of the aircraft and up in the tower in a flash. The instructor was not so lucky; he received burns to his hands, feet, and face. The dual student got out ok. All of us in the traffic pattern could see what was unfolding on the runway; however, nothing could be done to change the outcome. The instructor became the poster child for the military regarding the proper wearing of gloves, socks, and boots. Up till now, boots were not required while flying. One could fly with low-quarter shoes, no gloves, and any type of socks.

On one overcast chilly day, I was hovering up to the pad to take off when I noticed two people standing near the pad with some major camera equipment. I tried to hold the OH-23 very steady while they took their pictures. Several months later, Dad showed me the front page of the Chicago Tribune. It was me hovering the OH-23 at the launch pad. White socks and low quarter shoes. I liked flying with the low quarter shoes because I could feel the aircraft better. The shoes had a hard leather sole and transferred the vibrations and movement better to my feet.

VISUAL FLIGHT RULES (VFR) MANEUVERS *PARTIAL LISTING

- Take-offs (Normal, Maximum Performance)
- Landings (Normal, Steep)
- Normal Flight
- Slow Flight
- Slope Landing/Balancing
- Pinnacle Landing/Balancing
- Single Skid Slope and Pinnacle Landing
- Straight-in normal Auto-rotations
- 180-degree Precision zero ground run Auto-rotations
- Low-Level High Speed zero ground run Auto-rotations
- Hovering Auto-rotations
- Low-level (Nap of the Earth) Flight
- Water Landings (Theoretical)

We received instruction on how to approach confined areas and land. Now confined areas in Texas weren't bad. Scrubby little trees, snakes, and cactus were most of them. We were to land, throttle back to idle and friction the controls to prevent the aircraft from having a mind of its own. Then we were to get out of the helicopter and look around and find a rock that had a number or code written under it. We would then loosen the friction on the controls, increase the RPM by twisting the throttle, clear ourselves, and then take-off. Once having found the code then we were to continue to the next location. One of the students failed to get the friction tight enough and while out of the aircraft it took off without him. There was humor in that really. Imagine turning around and watching your only means of escape slowly rising and away from where you are standing. Then just imagine what the US Army is going to think of what you just did. What are they going to do, send you to Vietnam? Yep.

We flew around Possum Kingdom Lake quite a bit. Many of the confined areas were located around the lake. Our cross-country flights were always a real joy. I flew down around Abilene and ran low on fuel. Poor flight planning on my part. It was apparent that I didn't have enough fuel to get back to the Main Heliport. I landed next to a gas station and asked for a gas can. They looked at me a little strange. I went ahead and put about 5 gallons in the OH-23, cranked her back up again, and took off headed back to Ft. Wolters. It ran a little rough; however, it flew back just fine. Crazy stuff.

On this morning, we all fell into formation. The TAC Officer's voice was sad and then he grievously began to explain that one of our classmates had passed away. He had chosen this incident to teach us what true respect for the living and the dead should look like. He further explained that it was due in part to our careless recklessness that his death occurred. Some began to murmur, "who," "how," "when," the TAC then called out "ATTENTION". Then he carried on and said that due to the death all leaves and passes were cancelled this weekend. That the entire company would be at the funeral on Saturday. "Dismissed"! We all were talking, trying to figure out who was missing from the formation. One person said that he had heard that one of the classmates went down in a TH-55 and was killed on a solo mission. They were trying to identify the body. During the week, we were told that we were to be in our dress uniform for the funeral and that we would be marching to the cemetery. "He's going to be buried here?" Well, we had heard rumors that there was a cemetery on Post for students that had no homes. Saturday arrived, we fell out for formation. It was truly a magnificent sight to behold. Everyone all dressed up for a "parade" or in our case a funeral procession. The pall bearers were at the front of the formation and they led the way. The student commander called cadence as we marched to the cemetery. Upon our arrival and the positioning of the flag, guests and the formation, the cadre read the eulogy. "We are here to celebrate the life of Candidate Candle Fly. Whereas, Candidate Candle Fly being stationed at Ft. Wolters Texas and auspiciously a part of Warrant Officer Rotary Wing Flight class 66-11 was found to be deceased on 12 December 1967 at 0535 hours. Whereas, through the neglect of his fellow aviators, he was allowed to enter hostile airspace (light globe over the sink) and fly too close to the light. May he fly eternally in our memory and the wind be at his back!" Dismissed! Yes, these were the times when we learned true respect for living creatures. Instances such as these are what builds true character in the face of uncertainty. We were becoming pure tempered steel! No, we were becoming the pride of America.

On one day out solo, I was still a distance from the main stage field. I could see what appeared to be a wall of dirt rising to a height of about 500 feet. I could tell by the winds and turbulence that I was going to be in trouble if I couldn't get down on the ground fast. The aircraft was only rated for about a 20-knot crosswind component and that looked like a lot more than 20 knots. I was very fortunate to be able to land safely and park the aircraft. A dust storm in the U.S. is called a haboob. There I was about to be swallowed up by a haboob. I referred to those things as dust-storms.

My confidence was growing as I successfully accomplished more and more. It was, however, a struggle. The transformation to manhood was a painful experience. Often, I would be the cause of my own problems. We learn from our mistakes and mature accordingly. This type of highly motivated training forced you to try and exceed beyond your own expectations. It pushed you to your limits. There were times when you had to reach way down inside for motivation and the strength to keep on trying. The hazing was a daily endeavor by the TAC officers. For some of us, it meant reaching down inside to places that were not known before. This became the most important part of the training aspect for me.

Primary Flight School, Mineral Wells, TX, FT. Wolters

Pre-solo (hat is backwards)

Primary Flight Training, Fairchild-Hiller (OH-23D) "RAVEN"

3. "What Am I doing?"

FLIGHT SCHOOL ADVANCED FT. RUCKER, ALABAMA (20 YEARS OLD)

Drove from Mineral Wells, Texas to Montgomery, Alabama and spent the night. We came South down highway 231 when it was a two-lane road and drove across the Veterans Bridge North of Ozark. Talk about wilderness and back roads. We looked around and were astonished to find that we had gone back in time about 50 years. People still plowing fields using mules. It was a culture shock at an unbelievable level.

Welcome to Ft. Rucker Alabama, different sights, sounds, smells, and aircraft. It was a different world from the desert. It was lush and green, hot and humid and stormy.

APRIL 1966 — MEETING JOYCE

One evening after arriving at Ft. Rucker, Richard Morrison and I went to Troy to see if we could find a couple of girls to go out with and eat hamburgers or go to a movie. I mean all we ever saw were guys. Dick and I met two nice girls. After the date that night, Dick asked the girl (Sharon) out that I was with and I got stuck asking the other one out. Joyce was her name. Well, Joyce and I went out the next night double dating with Dick and Sharon. There was just something about this girl. I didn't want to date her, but it was as if I couldn't help myself. We would talk for hours and hours about our lives and what we wanted out of life. She went home to Dothan one weekend and I tagged along and met her mother. Now her Mother wasn't sure about me at all. Sometime in June, I started purchasing plates, cups, glasses, and a gravy bowl from the local gas station. I had lost my ever-loving mind. I know it was a "promotion" thing where when you bought 10 gallons you got a free plate and for .99 cents extra you received a cup with each fill-up. I knew that it was foolish, but I had to do something, and this seemed the right thing to do. I think I really lost my mind. Real high class thinking, I know. She and I had 12 dates during flight school. She gave me a reason for trying to get through and try to make the weekend without demerits. I would get off base on the weekend and go to Troy and just sit and talk or go grab a burger somewhere. Convincing Joyce that we were destined for each other was a very difficult task. She was a stubborn young lady. And a lady was just what she was when it came to me. We discussed everything from politics to how many children was the right number. We dated a couple of times at the Officers Lake Lodge (torn down 1997) at Ft. Rucker.

FT. RUCKER AL

- Advanced Flight Training—Ground School
- Instrument Training
- Blue Canoe (Instrument flight trainer)
- Bell TH-13T
- Instrument Flight
- Take-off (Normal and Maximum Performance)
- Landings (Normal and Steep)
- Special Instrument Flight Rules (IFR) Requests
- Normal IFR Flight Requests

INTO THE STORM

- Flight Plan Development and Filing
- Flight Following Requests
- Radio Procedures (Military Frequencies and use)
- Instrument Flight Rules (IFR) Navigation (Low-Altitude Enroute Charts)
- ILS, PAR, VOR, GCA, Tactical Non-Directional Beacons, Com Radios, DF steers
- Instrument Auto-rotations
- Water Landings
- Operation and transition in real-time
- Low Altitude Enroute Chart Navigation
- Visual Flight Rules (VFR) Sectional Navigation
- Dead Reckoning

ADVANCED AIRCRAFT UH-1A, B, C, D

- Night Flying/Navigation
- VFR
- IFR
- Inadvertent IFR
- Smoke and Haze (Now called IMC)
- Formation Flight
- Normal flight
- Night Emergency procedures
- Water Landings

ADVANCED FLIGHT TACTICS

- Landings to Minimum Lighting (Flashlight, Lighter, Smug-pots, etc)

- Single Ship Tactical Missions
- Multiple Ship Tactical Missions
- High-Speed Low-Level Flight
- Command and Control Missions
  - Coordination of tactical aircraft supporting ground troops
  - Directing Artillery Fire in Support of ground troops
  - Data gathering on road types, bridges, vehicles, troop movement
- Landing to Smoke (different colored smoke grenades)
- Proper Identification process
- Emergency Procedures
- Identification/resolution
- Security Operations Initiatives

Met the cadre and the senior WOC's, oh great now I'm working with sadists. Running was a way of life here. If you were outside of the building you'd better be running, or you were going to be doing push-ups for someone. If you weren't running, you'd better be dead. Hazing was a way of life here for the senior candidates. They had earned the right to make your life miserable and to make you quit if they could. If you couldn't take the punishment, abuse, psychological games, or physical pain, then quit. No one cared if you stayed or left. This was entirely volunteer and if you couldn't cut it no one wanted you around. The Army Aviation Community was known for a 40% washout rate in the Warrant Officer Rotary Wing Aviation Course (WORWAC). "I didn't make it this far to quit and I wasn't going to quit voluntarily." The hazing was very taxing, irritating, frustrating and in general continued the process of making you mean, insensitive, and totally uncaring about some things. Another incentive to not washing out was the fact that many of us would be reclassified to a military occupation specialty (MOS) of 11B (Infantry). For those that don't understand, that meant if you failed you had a ticket to Vietnam as a grunt, cannon fodder, expendable or a speed bump.

In all this frustration, we were dehumanizing. One was always struggling to find an anchor point. Many of us turned to relationships of the female persuasion. I was glad that I had Joyce to talk to and defuse my situational stress. She listened and was never critical; she would lend support when needed. She was my rock and my anchor in some troubling times.

Dating was one of those things that you tried to do just to get your mind off the negative aspects of the week. Between the hazing from the senior candidates, TAC Officers, military regulations, class workload, passing exams, polishing boots and buckles, and the flight instructors, it was truly more stress related than anything else. I'll never forget that almost immediately after arriving at Ft. Rucker they told us to Brasso the feathers off the eagle's chest on the USAAVNC crest. I say whaaaat!

More on being hazed and harassed by the senior candidates. I can remember one night they entered the barracks. We all hit the walls in the position of attention. Each time that we were addressed by the senior candidates or cadre we would answer in the following manner. "Sir, candidate Rawlinson (Yes or No), sir." On this occasion, the guy put his nose on mine and proceeded to shout at me and shout at me. Good grief I thought he was going to kiss me. The more he shouted the more I laughed at him. Finally, he ordered me to write a 5000-word essay on why I would not laugh at a senior candidate. Nothing was funnier than this guy in my face trying to be serious about stupid stuff. I just couldn't believe that this guy was serious. With what was facing us in the near future, I just didn't believe that he would make me do something like writing a paper. I still smile when I think of this incident. There's a time for fun and a time for serious—some never know the difference.

Flight training here was different. Since we already knew how to fly, we transitioned into a Bell TH-13T helicopter. We were to fly 50 hours in this aircraft, learning how to fly instruments. Instrument flying was broken into two segments; Basic and Advanced. This would allow us to fly under instrument flight rules (IFR) conditions. These types of conditions included blowing dust, smoke, rain showers, storms, fog, and cloud cover. It taught us how to navigate through these types of conditions and land safely at an airport or a tactical field location. Again, the Army's sense of humor reigns supreme. We were trained to the point of a standard instrument ticket; however, Warrant Officer Candidates didn't rate one. Instead, they issued us a "pink" ticket or a Rotary Wing Tactical Instrument card. I always tried to set my check ride for instruments on a real nice cloudy overcast day, so that we would have to fly an actual IFR mission. We would typically fly from Ft. Rucker to the Naval Air Station in Albany, Georgia.

Once we had mastered instrument flying, we moved on to Contact flying. It was during this phase of training that we transitioned into the UH-1 (Huey) helicopter. Joyce would help me study and, on this occasion, she memorized the start-up procedure for the UH-1 quicker than I did. I was excited as I approached the time to transition into the Huey, the UH-1, Iroquois a turbine powered aircraft. I was excited beyond belief. The basic flight characteristics were very similar to that of the previous aircraft that we flew. The differences were in size, power, governor, and feel of the aircraft. It was much larger and heavier than the previous aircraft we flew.

This became our work place, or our office if you will. This became our normal environment. Under the Airmobile concept we would:

- Climb and descend at 60 knots.
- Fly straight and level at 80 knots.
- In formation, we would fly with overlapped rotor blades.
- We performed formation approaches to the ground.
- We performed takeoffs both normal and maximum performance in formation from the ground.

- We learned how to perform backup 180-degree auto-rotations to confined areas.

Now the latter was hairy! The instructor cuts the throttle to simulate an engine failure. Student lowers to the collective. Locates a "suitable" landing area. The student begins a turn to set up the approach. Slipping the aircraft if necessary. Bleeding off airspeed makes adjustments to rotor RPM by adjusting the collective so as not to overspeed the rotor. Finds the spot on the ground, at 75 to 100 feet begin the deceleration, avoiding obstructions, wires, trees, continues to raise the nose to stop forward momentum, at 10 feet "POPs" the pitch to cushion and touches down with no forward movement. A piece of cake! Anyone can do it.

Airspeed and altitude control, if we were off by as little as 5± knots or 50± feet, one would get a pink slip and bust the flight. Too many of those and you were recycled or washed out of the program.

Survival training was here at Ft. Rucker near Lowe Field. The entire survival course was 7,000 meters long (about 4.5 miles). They made us strip so they could make sure that we weren't bringing any food with us into the field. They made sure that if we got hungry, we'd eat whatever they fed us. Some guys had stashed food in their pants legs. By the time they were done, they had a pile of food, snacks, candy, and other assorted items.

We were taken to a realistic Vietnamese village. It was as authentic as it gets. They assigned us to hooch's (little grass and stick shacks). When they came running into camp calling out loudly "Di-we, Di-we, VC come, VC come." We turned over the cooking pots and crawled into underground tunnels. They were quite long. We crawled on our hands and knees in tunnels to about 100 meters outside the camp. The course was tough enough. Couple of us that I'm aware of got roughed up pretty good on the Escape and Evasion course. This turned out to be my fault. We were in a group of three and we had been going through the thick undergrowth in the woods. It was tough going and we couldn't see well at all. Overcast skies, no moon at times, created an environment that was tough. We were getting torn apart by the undergrowth so when we came to a road and we took it. BIG MISTAKE! We got caught. I had an overzealous aggressor that caught me. He made me do 100 pushups in the mud while it was raining. I didn't mind that except that every once in a while, he would stand on my back as I tried to do them. I ate some serious mud that night. Three of us were captured and taken to a POW camp at Ech field. They had actual POW containment devices. Ok, enough with the realism, we get it. We escaped after overpowering a guard. I really think it was a setup to let us go and try and get back into running the course. We ran until we got into the swamp at the edge of the airfield and then we crouched down in the water near tree trunks. It was still dark, so we estimated we were about 200 feet into the swamp. The aggressors came to the edge of the swamp and were shining lights into the area and would occasionally shoot blanks towards where they thought we were hiding. We could see them, but they couldn't see us. We knew they wouldn't come to the swamp for us. We didn't dare sleep or move until they left. We had snakes crawling around us and we were sitting in about 18 inches or so of water. As the sun came up, we had a better look at our surroundings. Snakes here and there swimming around. Then we heard noise coming right at us. We just knew that the "aggressors" had found us. Then we see it; the biggest cat I've ever seen. About a 65-pound Bobcat come running right at us and it hadn't seen us yet. When he saw us, he put on the brakes, stopped, looked at the three of us and just walked away. Now that was enough for us, so we got our bearings; we headed back out on the course to finish the E&E course. The simulation was realistic. We were told that one had died during the E&E course and one was injured. One drowned and the other fell down an embankment and broke his arm.

The reward for the successful accomplishment of this task was fried grasshoppers, grub worms, and ants for lunch. They killed a rabbit and skinned him. They cooked and ate the rascal without sharing it with us. We got the boiled rattlesnake soup and meat from the pot.

INTO THE STORM

We moved from contact to the next phase of our training to Tactics. Tactics were all "green military". We lived in the field for a week and flew tactical missions both at day and night. We performed troop movements in flight formations. We shot approaches at night to various lightings. We landed to smudge pots, jeep lights, flashlights, and cigarette lighters. We performed artillery missions.

During this last little bit of training, we knew we had made it. We enjoyed this portion of the training most of all. Our time was coming to an end.

After nine months of absolute insanity, I looked back at where I started and was amazed at my growth as a person. We had all changed. We didn't talk the same. We didn't act the same. We had added a new lexicon to our vocabulary. A new manner of dress and decorum. More importantly, we had added a new skill set to our functionality. There we all sat in the theatre. I looked around at my fellow classmates and I was proud. I may not have a "tassel" to flip, but we just completed a course of instruction that four out of ten won't and don't. They told us at the beginning that this would be the equivalent to a four-year degree program concentrated into nine months. Years later when I received my undergraduate degree, there was no comparison to flight training in the Army.

Mother came down for my flight school graduation. I talked Joyce into coming home with me for two weeks. I had to promise to be a gentleman. So, Joyce, Mother, and I all piled into the station wagon and drove the 1,000 miles to Michigan. During those two weeks, I went to town and bought a modest engagement ring. I proposed to Joyce and to my sheer delight she accepted. I wasn't sure that she would accept since I would be leaving sometime soon. We had several nice days there at the Lake before Joyce left on a Greyhound bus bound for Washington, D.C. She was going to see her sister before going back to Alabama. A day later I was on a plane bound for Ft. Lewis, Washington.

Joyce (My Bride)

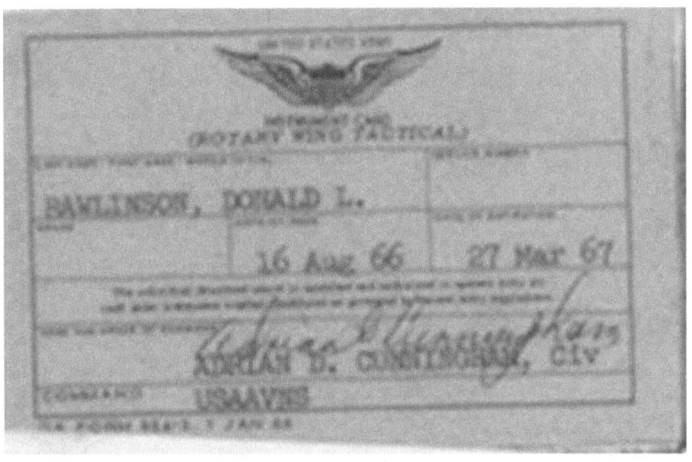

"Pink Ticket" (Tactical Instrument Ticket)

4. "I KNEW IT WOULD BE LIKE THIS!"

FT. LEWIS, WASHINGTON (FIGURATIVELY I CAN HEAR THE CANNON'S THUNDER AND SEE THE CANNON'S FLASH.)

Terry Langille and I arrived at Fort Lewis, Washington on about September 4, 1966. Base housing was unavailable at the time, so we decided that we would get an apartment in Tacoma. There was only a skeleton A Company organization and most of the officers and enlisted had not even begun to arrive yet. I still have the originally signed papers for the apartment. I'm a pack rat; what can I say. I suppose that's what made me pick up and continue to write. I'm going to get rid of it all. Some of the 4th Infantry Division was still on base and were moving out to head to Vietnam. I met one guy I went through basic with and we spoke for about an hour before he had to leave.

As the month passed quickly, many were showing up and the organization was beginning to take shape. Some of us noticed that a few of the commissioned officers were "old". I mean like my grandfather's age. The Commanding Officer MAJ John Willis had to be 65 or better. Then we met Captain McGowan, old as dirt, but wherever he went he went in style.

There were 19 of us from my class that were assigned together. We were some of the first ones to arrive at Ft. Lewis. (Company A inactivated 25 April 1966 in Vietnam; activated 1 September 1966 at Fort

Lewis, Washington).[1] We were here to reform A Company, 4th Aviation Battalion, 4th Infantry Division, callsign Blackjack.

We went in groups of five to Ft. Worth, Texas to pick up our UH-1 helicopters. I watched as 65-12895 came down the assembly line to be serviced and to perform its maiden flight.

Let's go over the flight we made from Ft. Worth, Texas to Ft. Lewis. How much could happen between just two places? A Huey could fly for about 3.5 hours on about 1200 pounds of fuel between 90 to 100 knots or 104 to 115 mph. A lead aircraft was selected for each leg of the route. It was the responsibility of the lead off aircraft to file the flight plan for the flight. Quick math, with no head wind we could make about 345 miles with a 30-minute reserve, in case of wind or bad weather. That meant that our flight path had to be created with specific fuel requirements, altitudes and hazards taken into consideration. The Iroquois or Huey, as it was affectionately known, carried no oxygen or parachutes so certain flight paths were not an option.

Our take-off, DFW(Dallas) -> Midland, TX (refuel) -> El Paso, TX -> Tucson, AZ (refuel) -> Yuma, NV. This was a troubling stop. The weather was deteriorating, we were off course to the North in the Mountains, limited visibility, tight formation in and out of clouds when the 20-minute fuel warning lights all started to come on. The lead aircraft was getting concerned that we might not make it and we had increased our speed to 120 knots. Well, this is when the little tiny hairs on the back of your neck start to lift. Your brain starts to say things like, "Great! Maybe I will get to do a marginal weather auto-rotation

---

[1] https://history.army.mil/html/forcestruc/lineages/branches/av/004av.htm

and try to do a slope landing on the side of a mountain." Then reality sets in and you begin to think. Is there any way to conserve fuel? Is that really a good idea? We need to get off this mountain range and quickly. It's flat on the other side. If need be, we can auto-rotate to flat land. The needle is still on the nose, the airfield is somewhere out in front of us, just how far?

Well we break out of the clouds and the airfield is there, and we shoot a straight in approach to land and shutdown. This is an important point, I no longer believe in "LUCK" anymore. There are simply too many things that happen that can't be related to "LUCK". While the aircraft are shutting down, running out of fuel, someone comes running out of operations. The Air Force had a student pilot crash into a butte. Their crash rescue helicopter was sitting there at the airfield with a flat tire. Looked like they hadn't flown the thing in a very long time. They asked if we would fly down and check out the crash site. However, the word radioed in was that the student never ejected. There was no point in checking out the site, so we refueled and continued our way. Sounds way too cold to be a human characteristic, however, now it's a thought process carried out using physics.

Next stop is sunny Palm Springs, California. Good grief, where did all these mountains come from. We've been climbing for a while and what the heck are we breathing. It certainly isn't air. We're over 10,000 feet, eyes burning, lungs burning, the air is a dirty dingy looking color. I'm asking on the radio what is it? Someone responds with "smog". It was my first experience with smog. We begin our descent into Palm Springs. The lead aircraft radios the tower that we're a flight of five Army helicopters, full stop and refuel JP-4. Then he follows with the comment/request for special hovering instructions. The Huey's rotor wash can be up to 100 mph, therefore, if we get too close to non-tied down aircraft, they could be damaged. Our warning went unanswered, in we came and transferred to the ground frequency and issued the warning again. They hovered us right down into the taxiway and into the parking area while the rotor wash played havoc with some of the aircraft. A little while later we hear "what's hovering and rotor wash?"

Refueled and ready to go again, we begin the scenic route North. If you look at a map one might see that it was scenic. We followed the lead aircraft, sometimes wondering where he was going. We flew NNE toward the Mojave then W to Barstow WNW to Bakersfield NNW to Stockton, CA. This is where we remained overnight (RON). It's about 9 PM (civilian time) or 2100 hours (military), we rented a car, piled in, and head to San Francisco. Well, what can I say, it's an Army "thing". I can't remember if we got any sleep that night or not. As we flew out in the morning, we low level (altitude 50 feet above the ground) to the mothballed fleet in Suisan Bay. We continued through Redding and prepared for the mountain ranges ahead. We negotiated our way through the valleys of Mt. Shasta, targeting Klamath Falls, Grants Pass, Eugene, Mt Hood, Portland, Olympia, and finally Ft. Lewis, WA. I was told that of the 25 total aircraft that were picked up, one Huey was lost in the Mojave (engine failure).

Once we had our full complement of UH-1's (25 aircraft). The Commissioned Officers had us fly to Yakima on several occasions to pick up full aircraft loads of apples. Beats me what they did with all of them. Entrepreneur's now that I look back at what they probably did with the apples.

Once we got our aircraft to Gray Army Airfield (GAA), I flew every chance I could get. I was one of the original "stick pigs". What this meant was, if you didn't want to fly, I would fly all day long. All of us needed to fly whether we wanted to or not. I flew every chance I could. The Huey is a two-pilot aircraft according to the Army; however, it can be flown by one person. By adjusting the weight and balance (putting a 150-pound weight in the left seat) one pilot could fly it perfectly. In fact, it flew even better with just one person in it. So just where did I go one might, you might ask? Here are the locations that I went and what

INTO THE STORM

I did. Remember throughout flight training we performed every one of these tasks solo. So, I went cross country to Mount Rainier, Snoqualmie Pass, White Pass, Yakima Orchards, Yakima Range, Sunrise and hovered in places where only Angels flew. I flew low level down the side of the mountain at 130 miles an hour leveling off at tree-top level and twisted and turned through the valleys as only an eagle could. I have been to SEA-TAC, Puget Sound, Naval Air Station Whidbey Island. I flew out over the cascades and landed on the shore, shutdown at a town by the name of Seaside. I landed on the beach on the Pacific, shutdown the aircraft and went swimming in the Pacific. I showed off the Huey to those that came by while I was sunbathing on the beach. I landed at a restaurant on the coast and ate Dungeness crab.

One time, I can't remember who was with me, but we came down the mountain low-level and shot out across a clearing and suddenly, a huge chicken house appeared. We were just off the top of it. Well, we didn't think any more about it. Upon arriving back at the base, we were "asked" to report to the base commander's office. Totally in the dark as to what we may have done we drew a blank. Neither I nor he was prepared for the butt chewing that we were about to receive for killing as many turkey's as we did by pulling a stupid stunt over the turkey house. He let us know exactly the cost of the turkeys. We suggested that since Thanksgiving was just a few weeks away that perhaps we could salvage some of the cost. He didn't find any humor in that since it simply set him off again. Wisdom is knowing when to speak and when not.

The pilots received qualification on the M60 machine gun on the Yakima Range. At this point everyone is saying, you guys had it made, look at all the fun you're having. Yep, and I was sucking it up and making the most of it while not talking about any of it.

Then while some were gone to pick up more Huey's, this happened..

Date: 1 Oct 1966
West Coast Airline DC-9

Wemme, Oregon
N9101
Fatalities: 18 of 18
https://en.wikipedia.org/wiki/West_Coast_Airlines_Flight_956

West Coast Airlines Flight 956 crashed into Squaw Mountain approximately five miles south of Wemme, Oregon. All aboard, crew and passengers perished. The NTSB requested Sixth Army to assist in the recovery of bodies and black boxes. Since we had the only Army helicopters west of the Mississippi, we were called upon to aid in the task. I will never forget the crash site. It is enough to say that it was devastating. The aircraft disintegrated into pieces upon impact and burned on the mountain and down the side. John Morgan from our unit was flying with a Sixth Army Major Rogers. With a full load of NTSB officials, they lost control of one of our newly acquired UH-1D models and rolled it down the side of the mountain until it came to a stop inverted minus the engine, part of the tail boom, crushed cabin and a tree stopped by John's armor seat and his head. John was in Madigan Hospital for two weeks suffering from amnesia.

As a note, Van's Aircraft Inc. employee John Morgan and his passenger, Lawrence Hull, were killed May 24th, 1998 in the crash of the second prototype RV-8, N58RV. The accident occurred near Blythe, CA.

We were crazy—certifiably nuts. No one would believe the crazy stuff we did and lived to tell the tale. Then again who would believe it so why talk about it. A few weeks after the picture, I decided I would like to fly up to the same spot and look around. It was so cool to sit in the seat and be only feet away from where I stood weeks earlier.

Occasionally we would be asked to take some of the folks on the beach for a ride. I loved flying over the Cascade Mountains and up Puget Sound. The weather was normally always bad and challenging, but the days that were nice more than made up for the bad weather.

We would fly into White Pass and Snoqualmie Pass around Mt.

INTO THE STORM

Rainer and go IFR (in the clouds) and then have to find a hole to climb through to get across the range to Yakima. Sometimes we would hover up to the mountain and then pull pitch to climb vertically. (We were crazy and what were they going to do? Send us to Vietnam!) We had developed an attitude of invincibility.

I loved flying up to Sunrise and Paradise Parks. Both locations gave totally different perspectives of the mountain. "The steam that vents from the remnants of a dominant volcano still exists today." This type of mountainous flying really helped to prepare me for the weather that I would encounter on several missions in Vietnam. I would assume that if one is meant to fly, they can even defy physical laws and some scientific ones as well.

Total flight time accrued, 175 hours in and around the Cascades, Yakima, and Mt. Rainier.

We creatively appropriated a 50KW generator from the Navy Air Station junkyard located in Puget Sound to take to Vietnam. The Navy is probably still looking for that thing.

Fourth Infantry Division Patch—Ivy Division

DONALD RAWLINSON

Indo-China war

1. French went into VN in strength in 1946.

2. U.S. began giving extensive economic and military aid to French in Sept. 1951.

Gulf of Tonkin

3. Aug. 4, 1964 both the Maddox and the destroyer C.J. Turner Joy were attacked. In retalliation Pres. Johnson sent U.S. planes to bomb naval installations in North VN.

4. In Feb. 1965 VC attacked an air base near Pleiku. Pres. Johnson sent planes to bomb barracks and staging areas in North VN.

5. In March 1965 Johnson ordered troops into South VN.

6. In Feb. of 65 US troops amounted to 24,000 troops at the end of 1965 there 215,000 troops in VN counting 40,000 naval forces.

Page from training we taught prior to shipping out

INTO THE STORM

My perch on Mt. Rainer overlooking one of the glaciers—Oct 66

DONALD RAWLINSON

Me qualifying on the M-60 machine gun. Yakima, WA

4th Combat Aviation Battalion Crest

Joyce and I leaving the wedding under crossed sabers — Nov 26, 1966

Joyce and discussing why I can't cut the cake

5. Are Marriage and the Army Compatible?

IT HAS BEEN SAID THAT "IF THE ARMY HAD WANTED YOU TO HAVE A WIFE, THEY would have issued you one." I'm sitting on the airfield thinking, I sure miss Joyce. There's no doubt in my mind I want to get married. Really, I don't want to wait. Then I go, "Why don't we get married before I ship out?" She will have more money for college, be able to shop at Ft. Rucker, and if I die, she gets the insurance and benefits. Ok, I know that I'm trying to justify it, so I can make a real positive sounding argument for the case of marriage. Now look closely at some of my points. What looks positive to a guy may not be positive or real solid reasons to a woman. Guys are logical, women not so much. I began to burn up the phone lines to convince Joyce that I had good points and that we should get married now. We would even have about six weeks before I shipped out. Reluctant wasn't the word. Finally, she said ok.

What's this? I've got to get my Company Commander's permission to get married and to top it off, Joyce must be interviewed by the Chaplain. The roadblocks were coming so fast and I thought Joyce was going to be hard. MAJ Willis listened to my plea, then simply said, "Why now?" My reply was just as simple, "Because I love her."

Chaplain Lamb said he doesn't marry people he hasn't interviewed. He didn't negotiate. Finally, I asked, "Would you consider interviewing her over the phone?" He replied, "Yes". Finally, a thumbs up. Now we need to do what? Marriage licenses—one from the state of Washington and one from Uncle Sam (sugar). Man, this is crazy, all I want is a Justice of the Peace.

This man's Army wanted a full military wedding. Eight days of insanity flourished on Ft. Lewis as I tried to keep my mind on work and everything else. Richard Morrison (a great friend) helped to keep me on course. Joyce and her mother arrived at SEA-TAC and I was there to pick them up. I watched diligently as the passengers disembarked. I didn't see Joyce or her Mother. Now I was nervous—did she back out or miss the flight?

    I was looking all over the place when I heard her mother's voice call her name. I turned around and there they were, looking for me. They had the same thought. I must've backed out and now they were 3,000 miles from home. Are all weddings so stressful?

    The day of the wedding finally arrived. It was November 26, 1966, 1600 hours. We walked out from the back into the church and the pews were filled with soldiers there to witness our marriage. Everyone was in their dress blues; it was magnificent. We all took our places, the music began, and I turned to look toward the entrance.

There representing Joyce's father stood my Commander, MAJ Willis, with my bride on his arm. In just twenty-six minutes we were married and walking down the aisle. Now I ask you, shouldn't the ceremony that takes weeks to plan, multitudes of people doing all sorts of things, and the marriage supposed to last a lifetime last longer than 26 minutes? A man's view or take on the subject.

    They had booked two rooms at the Holiday Inn for the night for us. One room was for us to get adjusted to one another and the other to go to bed. Then they gave out the wrong room number to where we were. Now the poor guy in the first-floor room that got harassed all night didn't appreciate that one bit. It was dawn when they came to get me.

We were starting a two-week field exercise in the mountains. Man, I just got married a couple of hours ago. Hey, if the Army had wanted you to have a wife, they would have ISSUED you one. That was the response that I got in the morning. In some twisted way, it made sense to me.

Married Helen Joyce Dukes on November 26, 1966, and at precisely 1626, I said, "I do." Joyce's mother, Thelma, came with her to the wedding. No one attended from my immediate family, but there were over 200 there from my military family. Joyce and I lived in an efficiency apartment for about six weeks while I prepared to go to Vietnam.

The following morning after we got married, I had to leave for a two-week FTX (field exercise). Short honeymoon wasn't it. We left from Gray Army Airfield. I remember a flight of six of us. We were to rendezvous with our force somewhere to the south towards Oregon. When we got there the weather was wet (rainy) and cold.

This is interesting. We're miles from nowhere. We've left the aircraft to pitch our pup tents. Now even the word "pup" should tell someone how big this stupid thing is, and they expect two of us to sleep in one. Oh, we get to have our own, that's great. Ok, let's inventory what we have: four small poles and two pieces of canvas that button together and some "string". Now have I missed something along the way. I've been in the Army for 16 months and I have been shown many things. Guess what, putting up a pup tent wasn't one of them. Number 1, pick a spot—ok does it matter where I put it? Oh, yes it does. Two, there's no floor, how do I keep the rain out? Trench around the base of the tent and the water will run off and right into it. No problem. Three, layout the two pieces and button them together. Again, how does this help keep the water out? Four, insert poles into each other and into the tent and fasten with string. Five, raise the tent and use tent pegs to tighten tent. Six, tighten strings and trench around the base of the tent. I get up and look at my work and I'm quite pleased with myself. I put my sleeping bag inside and I have an instant home away from home.

It was so blasted cold. I put a candle into a small C-ration can and wired it to the tent post. Just the heat from the candle made it warmer in the tent. We flew missions in the rain and snow. Just where is Vietnam again? After about three days I had pneumonia. We took off on a mission. There were three aircraft involved in the sortie. As we were coming back in to the base camp, we could see that the hole in the clouds was closing in. I decided that if I couldn't see them then I could return to Gray field and go home. Well, I radioed and said that we were no longer able to continue approach in to the base camp. They were not happy campers. I turned and took up a heading to Gray Field.

The weather was getting so bad that we were being driven towards the ground. It was becoming difficult even to maintain contact with the ground. Gray Field was reporting zero-zero conditions, that meant that you had no visibility to land. Well, I had enough cold and bad weather. I saw a road that I was familiar with and I went down and landed. I sat there a few minutes to make sure that I knew where I was. We weighed what our chances were to get to the airfield. Questions like were there power lines across the road. Now, this is true IFR flying (I Follow Roads) that is. We discussed it and we were all of the same opinions. I pulled pitch and began hovering up the road. I went a little way and once I recognized where the fence was for the end of the runway, I jumped the fence. I called the tower and they asked our location. I radioed that we were on short final for the runway. They said that they couldn't see or hear us yet. I radioed back that we had the runway in sight and should be touching down in about one minute. I hovered until I found the runway and then we started feeling our way back to the tie down area. It was during this time the tower radioed back and said they could hear me but couldn't see me. They said that the tower was socked-in. I radioed back and assured them that we were on the ground hovering to the tie down area. Once we got close, the crew chief got out and walked directly in front of us so that we wouldn't hit anything. We finally got to where the parking area was and shut the aircraft down. I heard about that episode for months.

Joyce took a job cleaning living quarters (I didn't know that she did it.) I was devastated to think that I couldn't support my wife without her having to work. We shoved a lot of living into six weeks.

CW3 Dunaway and I flying in support of the 2 of the 35th March 2nd, 1967

65-12895 after we were shotdown on April 1, 1967

The Air Force spraying Agent Orange

B-52 performing a Arclight mission

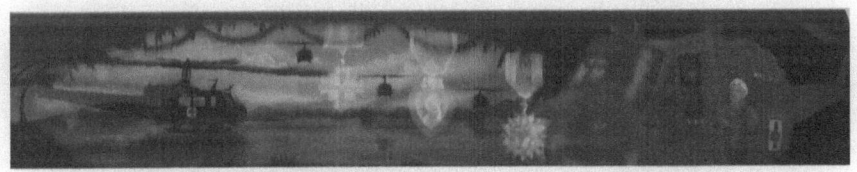

My retirement Huey tail rotor blade (painted by John Wayne)

CPT Recher standing next to a badly damaged UH-1D
Vertical fall into a stovepipe landing zone (LZ)

A draft card burner—also someone that caused Ho Chi Minh to continue to fight and not surrender

Charles "Chip" Pfordt and Robert Nash working some "furniture"

Best of friends Roman Ochotsky and Richard Morrison

A picture of the "Golf Course" at An Khe 1st Cavalry Division Airmobile

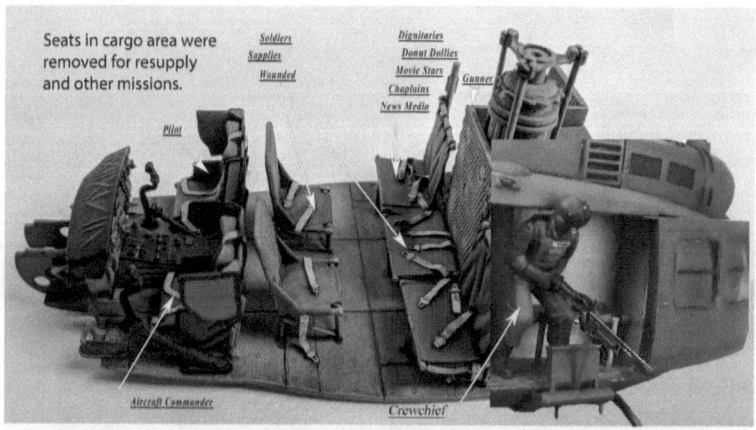

Model of floor plan for UH-1—Typically there were no seats in the cargo area

My 1-dollar MPC with Jayne Mansfield's autograph

Joyce sent me this picture of her

A picture I sent to Joyce

A picture of the left foot pedal—still has the blood stains on it.

INTO THE STORM

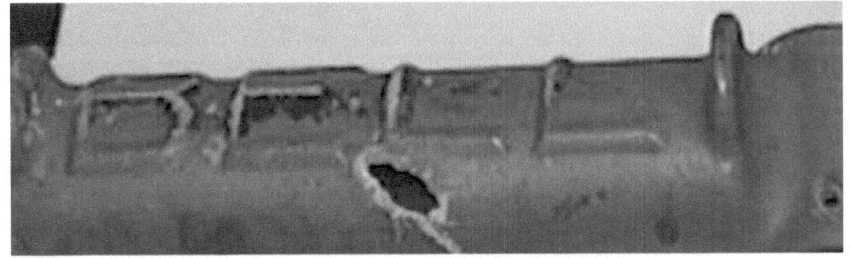

You can see some of the tar embedded in the "E"
probably has some Agent Orange in it still.

Got to see Nancy Sinatra

Took off and snapped a picture of Siagon

A1-E Skyraiders (Sandies)

INTO THE STORM

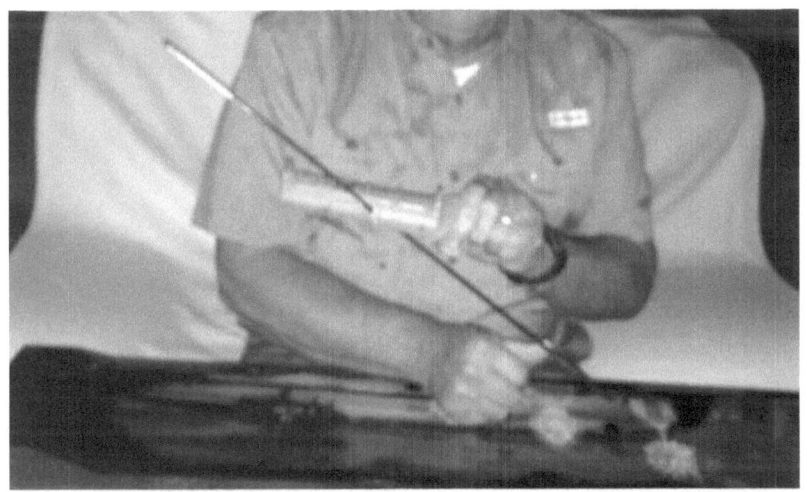

Bullet trajectory through the pedal

Where the VC loved to hide

DONALD RAWLINSON

```
650A CST APR 5 67 NSA026
SPA014 PA710 P WA016 XV GOVT PD CAS FAX WASHINGTON DC
5 1226A EST
MRS HELEN RAWLINSON, DONT PHONE DONT DEL BTWN 10PM AND 6 AM
   1006 MEADOW LANE OR DOTHAN ALA
THE SECRETARY OF THE ARMY HAS ASKED ME TO INFORM YOU THAT YOUR
HUSBAND WARRANT OFFICER DONALD L RAWLINSON WAS SLIGHTLY WOUNDED
IN VIETNAM ON 1 APRIL 1967 AS A RESULT OF HOSTILE ACTION. HE
SUSTAINED A GUNSHOT WOUND TO THE LEFT FOOT. HE WAS THE PILOT
OF A HELICOPTER LEAVING THE LANDING ZONE WHEN AIRCRAFT WAS
HIT BY HOSTILE SMALL ARMS GROUND FIRE. HE WAS TREATED AT 18TH
SURGICAL HOSPITAL APO SAN FRANCISCO 96318 AND IS BEING HELD
FOR FURTHER TREATMENT. ADDRESS MAIL TO HIM AT ABOVE NAMED MEDICAL
FACILITY. SINCE HE IS NOT REPEAT NOT SERIOUSLY WOUNDED NO FURTHER
REPORTS WILL BE FURNISHED
   KENNETH G WICKHAM, MAJOR GENERAL   USA THE ADJUTANT GENERAL
```

Western Union Telegram Joyce received.

6. "Saying Goodbye?"

WELL, IT WAS TIME TO LEAVE FT. LEWIS AND MAKE OUR FINAL PREPARATIONS TO be deployed.

Joyce and I rode the Western Flyer in Vista-Dome cars (a train from Yakima to Chicago). It took 3 1/2 days to reach Kalamazoo; I had a stateroom. However, someone boarded that had more clout than me. Now we had berths across the aisle from one another. We would hold hands across the aisle. We came via Washington, Montana, Utah, Wyoming, North Dakota, Minnesota, Wisconsin, Illinois, Indiana, and Michigan.

Joyce taught me to play strip poker; she always won. Imagine that, naïve wasn't I; well my excuse is that the Army neglected to work with us on naiveté.

There was an older woman on the train that helped us out by buying our lunches. Remember I was an officer making a lot of money (sarcasm intended). She rode the train to visit her children. Neither wanted her so she just rode the train back and forth to each of their respective households. Joyce and I felt so sorry for her.

Once in Michigan, we visited the family. We then had to find a way home back to Alabama. We were stranded in Michigan. Joyce called her mother and we borrowed $800 to buy a 1962 black VW with a gas-fired heater. I loved Thelma for bailing us out like that. I really felt indebted to her for the rest of her life.

Joyce had to buy it because I was too young to own any personal property in Michigan. I could not believe that not only could I not drink a beer, now I wasn't old enough to get married or purchase my own car. How is it that I can join the Army, fly a quarter of a million-dollar aircraft and go to Vietnam but not own any personal property? Second thoughts anyone, anyone.

Joyce and I passed through Indianapolis during one of the worst ice storms in history, December 1966, on the way back home. Joyce and I spent what time we had remaining together in Dothan.

I can remember like it was yesterday the day Joyce put me on the airplane in Dothan. Joyce handed me a letter and said not to read it until I was in the air. The reality of what was about to happen began to sink in.

Saying goodbye hurt—but it was time to leave. I had to really work at not thinking about her the next few days. I knew that I wouldn't be able to focus on final preparations to deploy.

Once I arrived back at Ft. Lewis, we only had a couple of days before we were to pull out. I said goodbye to the friends that we had made in the short time that we lived there. Then it was time to go.

8 JANUARY 1967

We got ready—loaded on the bus and went to McCord Air Force Base located to the North. We boarded a highly-polished aluminum skinned C-141 Starlifter. We flew to the North into Alaska and landed for fuel and then we waited, for what we didn't know. Finally, we left there and flew to Japan, and landed. They kept us on the plane for hours. Seems that no one planned very well, and we would have landed in Pleiku during the night. Since it would have been really easy to shoot us down, someone elected to make us wait until we could land during the daytime. The rumor was that the airfield at Camp Holloway came under fire and they couldn't guarantee a safe landing.

INTO THE STORM

The Starlifter became a casualty and had a short life in the country.

>From: Difante Archangelo Civ AFHRA [Archie.Difante@maxwell.af.mil]
>Sent: Monday, September 25, 2000 3:47 PM
>To: donald.rawlinson@se.amedd.army.mil
>Subject: C-141A, s/n 65-9407

Dear Mr. Rawlinson

Thank you for your request. The aircraft in question had a very short service life. It was gained into USAF service on 21 Sep 1966 and was assigned to the 62nd Military Airlift Wing (Military Air Transport Command),

McChord AFB, WA. It was deployed to Da Nang AB Vietnam and assigned to the 366th Tactical Fighter Wing (Pacific Air Forces) on 22 Mar 1967. It was dropped from inventory due to a flying accident on 14 Jul 1967.

I hope this helps.

Sincerely
Archie DiFante
Archivist
(334) 953-2447
AFHRA/RSA
600 Chennault Circle
Maxwell AFB AL 36112-6424 USA

This was a tribute to a fallen aircraft that died not as a result of war but an accident.

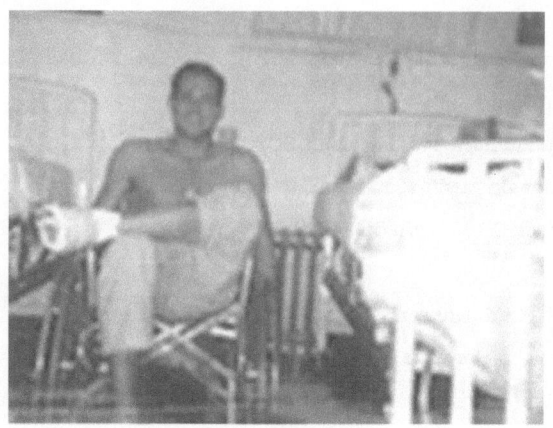

Johnson AFB Japan

Johnson AFB Japan

INTO THE STORM

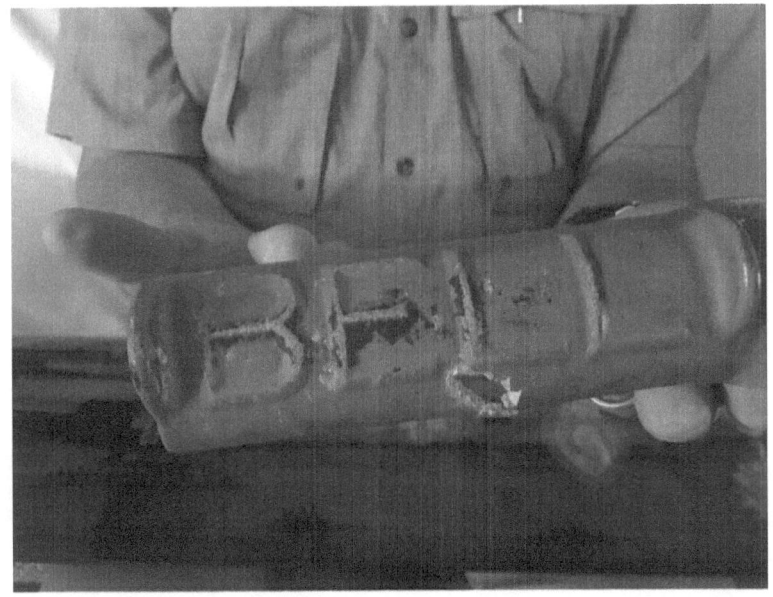

Pedal

7. War or Vietnam Conflict

REORGANIZED AND DEPLOYED A COMPANY 4TH AVIATION BATTALION (9 SEP 66 — APRIL 67)

4TH INFANTRY DIVISION, 4TH AVIATION BATTALION, PLEIKU VIETNAM (DADDY WHAT DID YOU DO IN THE WAR?)

I served in the following capacities:

Airmobile Concept—Combat troop insertion—put the troops in the LZ (landing zone). Slicks, as they called us, because we were naked or minimally armed, (2 M-60 machine guns).
Forward Artillery Observer/Controller—directed artillery on the enemy's position.
Reconnaissance and information gathering—roads, trails, bridges, troop concentrations, body counts
Command and control—Fly high to direct missions, control aircraft spacing
LRRP (Long Range Recon Patrols)—insertion and extractions
Flare drops—night duty during attacks. Front row seat to watch the tracer fire.

Night missions were conducted, however, were discouraged due to how dangerous they were.

Health Control Officer—Pass out prophylactics for weekend leave
Vector Control Officer—Rodent control and pest control
230 plus combat flight hours resulting in the following: (Who cares category.)
Purple Heart—wounded in action on April 1, 1967
Distinguished Flying Cross—March 22, 1967
Four Air medals (100 hours of combat assaults)—Over the course of time.
Vietnamese Cross of Gallantry w/ Palm

Aircraft tail numbers that I flew:

64-13705, 809, 835, 829, 831, 833, 65-9575, 65-12895, 65-12894, 65-12893, 65-12889, 66-748, 66-747, 66-775, 66-750, 66-751, 66-749 (Reflected in my log book does not reflect multiple ship flights on same day due to maintenance, accidents, combat disabling events.)

Units that I supported:

26th ROK Regiment, Tiger Division, and the White Horse Division, 1st Division, 2nd and 3rd Battalions, 1st of the 22nd, 1st of the 8th, 1st Cavalry, 2nd of the 35th, other units known only to God.

People in the 4th CAB I remember flying with:

1LT Sid Richardson, Lt. Col. Jerry Orr, CW2 Bernard Paez, Cpt. Recher, Maj. Duane DeBoer there were more, I just can't remember who they were.

INTO THE STORM

Friends with me:

WO Terry Langille, WO Quiett, WO Walt Melson, WO David Peters, WO Chip Pfordt, WO Richard Morrison, WO Robert Nash, WO Ray Pollock., WO John Morgan., WO Jim Moore.

9 JANUARY 1967

Finally, we made it. We were, in fact, standing on the tarmac at Camp Holloway, Vietnam just outside the city of Pleiku. I reflected for a moment about walking off the C-141 where they handed me six rounds of .38 caliber ammunition. Count them six whole rounds of ammunition, wow going to be a short war. Then as I walked a little away from the C-141, I turned and looked at it in the morning sun. She sparkled in the morning sun. She too was brand new and now in Vietnam as well.

We're all standing around when several armed A1-E's (Sandies) were getting ready to takeoff. They were impressive: big radial gas burning engines, World War II birds perfect for this war. They were loaded with Napalm and what looked like 200-pound bombs. As they took off one experienced an engine failure and headed for a ravine off to our left. We were all standing there watching as he crashed down in the ravine. The aircraft came apart in big chunks. Good Morning Vietnam and welcome.

We board a Chinook. Now I had never flown in a Chinook before, I wasn't worried until I noticed the crewchief standing under the rear transmission watching the transmission fluid leaking out onto the floor. I'm thinking this can't be a good thing. We are off and flying toward Dragon Mountain, home of the Fourth Infantry Division. It was about 11 kilometers to the South.

Our new home turned out to be a huge plot of red dirt with nothing on it. So, we begin to set up base camp. We walked into a field and someone said this is home for the next 12 months. Later we set-up GP medium tents, I believe there were eight or so of us in a tent. Everyone staked out some space and began building some wood floors, tables, desks, and bookcases. It was amazing how creative everyone was. Especially building out of shipping pallets. The old guys that were with us knew the ropes and lived higher on the cow than the rest of us. They'd been through it before and now they were doing it again. Boy, they knew how to live. Within a month, we had hot showers. The Engineers got together and built a big shower system. They used two Conex containers and plumbed the showers. It was great. Another group built some group outhouses. Oh, now that's a story all together. What does an Army outhouse look like and who pulls what duty? You can't imagine the waste burning detail. The smell brands you forever.

A few days later, we're all divided up and sent out to units for in-country training. We were flying with pilots that had been there at least six months.

Notes from my diary, logbook, and letters to my wife.

LEGEND

Date of Action — Jan xx
Aircraft Type — UH-1D
Aircraft Tail number — 64-13705
Number of Landings — Lnds 27
Type of Mission or Location — Resupply
In support of (Unit) — 26th ROK Reg
Type of Landing (DCS/CA) —
Number of flight hours — 5:20

January 10, 1967

I took some pictures of the wide-open spaces and red dirt. As I slowly turned around, I noticed tubes scattered around the area sticking out of the ground at an angle? Now, what purpose do you suppose they served? Off in the distance, I saw several people standing over some pots burning stuff, now what do you suppose that's about? I saw dirt, tents, and not very much of anything else. They've already got the enlisted guys filling sandbags. If we needed sandbags, wouldn't they already be filled? There is so much here I don't understand already. I'm sure it will all make sense in a short time.

January 11, 1967

The day passed uneventfully until after dark, then we could hear the mortars and 155's in the distance. About 1300, I leaned against a sandbag bunker and the wall fell. A real confidence builder. We all had cold beer tonight and lots of it. You must love the Army, soldiers are so easy to please. Just give us beer. I didn't smoke, drink, or cuss when I joined the Army. What you find is that cussing begins immediately upon getting off the bus for basic, smoking begins after the first 50 minutes of training and "if you've got-em light-em," and beer drinking begins immediately after falling out in the evening. I can't wait to see what waits for me here. Someone said that the big guns will have you drinking in a couple of days just to make you go to sleep. Ok, so at this moment I'm fixed on sounds of big guns and drinking. I thought I was here to fly? Silly Warrant Officer, you just don't get it yet.

January 12, 1967

Briefing all afternoon on orientation in V.N. I had a hot shower and watched a movie. I've got sunburn. What did I learn, you might say? Don't get shot and don't get captured. Don't fly too far west. Our call sign would be "Blackjack" + last three numbers from the tail. Attitude check!

January 13, 1967

Tents arrived, and we worked all day unloading equipment and putting up the tents. We are supposed to go to different units for in country orientation. I'll be with the 161st AVN Co., 14th AVN BN, Callsign: Pelican located at Lane Army Airfield—Qui Nhon. As I understand it, we'll be supporting the Korean ROKS: Tiger Division and the White Horse Division.

Welcome to downtown dirt hill, the accommodations are the following: Tent, cot, sleeping bag and plenty of dirt for the floor. Well, my education is constantly expanding. The tubes in the ground are like urinals. We're NOT supposed to pee on the ground. We're supposed to pee in the "tube". Not a problem, I'm civilized, I can pee in a tube in the middle of an empty field. No sweat. Now the content of the "pot" those soldiers were stirring, and burning is wwhaaatttt???!!! Ok, we have outhouses: 1) the pot is a 55 gal drum cut in half, 2) it's placed under the seat, 3) your business goes into the "pot", 4) The pot is removed/replaced with a "clean" pot, 5) Soldier: a) takes pot, b) adds jet fuel, c) stirs mixture to a even consistency, d) ignites the JP-4, e) stirs continuously until all of the contents are consumed by fire. Ok, of all the job descriptions in the Army, I'm curious, is there really a job like this in civilian life? Ok, I'm "outta" here this stuff really smells bad, I pity the guy that must burn that stuff.

January 14, 1967 (Sat)

The front gate received fire last night. I wonder should that bother me. How far are we from the front gate? It looks though we'll be moving tonight or tomorrow morning. I borrowed $20 from Cpt. McGowan for a radio. [I wonder if I still owe him the $20.] Note: I (2019) still have the radio and it still works. Getting bored, agitated, frustrated, I'll write a letter.

January 15, 1967

Back on a Chinook, the crew chief is standing under the rear transmission watching the oil drip out. Is there a problem with the transmission? We're moving to Qui Nhon today for our 10-day in country training. We are flying with the 161st 14th Aviation Battalion. Upon landing, we attend a three-hour briefing. Told us things like, "You'll get shot down if you fly too far west (Cambodia)." "Don't mess with the Koreans, they will kill you." "Don't eat the monkey meat, it'll give you worms." Wait a minute, what Koreans? There are Koreans in Vietnam, why? There are two Divisions of Koreans in Vietnam, the "White Horse" and the "Tiger". What is a "ROK"? It's what you throw at a Wabbit. I thought it was funny at the time. (ROK = Republic of Korea) I have to say that while in flight school the common joke was, "What do you throw at a "Wabbit", ans. A WOC or "Warrant Officer Candidate". The excitement is building, I can feel anxiety building in my gut. Or maybe what I think is anxiety is just plain fear. No, it's just excitement disguised as anxiety. We're supposed to begin flying tomorrow.

January 16, 1967

> UH-1D
> 64-13705
> Lnds 27
> Resupply Lane
> AHP 26th ROK Reg
> 5:20

I am flying with an old-school mate (Neil Elliot). Neal was sent to Vietnam directly from flight school. He's made aircraft commander and he's become, one fine pilot. He let me fly almost the entire time while narrating what to do next and how to do it. On the way out of one of the landing zones, I heard a noise I had never heard before. " Neil, what's that noise?" "Oh, don't worry about it. We're being shot at." He goes on to state, "That's what machine gun fire sounds like." "Oh! Should I be worried?" "Only if they hit us." "What's that sound?"

"They hit us in the rotor system." "Oh, should I be worried?" "Nope." "Hey Neil, what is that?" "Captured arms and ammunitions, happens all the time." "Hey Neil, what's that big pile of white stuff?" "Opium they capture it by the truck load." "What are we doing for the rest of the day?" "Just resupplying, "Ash and Trash", the troops and taking care of them." "Neil, do you ever get excited about anything anymore." "Not in months, how about you?" I responded not in the last few seconds.

January 17, 1967

> UH-1D
> 64-13829
> Lnds 21
> Resupply Lane
> AHP 26th Reg
> 3:20

We flew 3 hours 20 minutes on resupply for the 26th ROK Regiment (Korean Killers). The rest of the afternoon I had off. I've caught one heck of a head cold. It sure is cold and rainy. We flew into some really bad weather. We just about ran out of gas. Landed in Qui Nhon with 30 pounds of fuel. We hauled 10 Viet Cong back to the base camp. We had no problems. I wonder now who had them covered? Just kidding, the crew chief and the gunner had them under cover.

SPECIAL PEOPLE

Who the heck is Chris Noel? What is she doing here? Who put their career into a holding pattern at the beginning of a promising career? Why, did she do that? (https://www.facebook.com/chrisnoelactress)
    Chris Noel (born July 2, 1941) is a California actress of the 1960s.

INTO THE STORM

She is best known by veterans of the Vietnam war for her work on the Armed Forces Radio and Television Service. Her program "A Date With Chris" is fondly remembered as "the voice of a California dream girl" by many Vietnam vets. Her pin up pictures were very popular.

She is an author, radio celebrity, movie star, a representative for veterans. She has suffered from many of the same trauma's that combat veterans have. She has been shot at, shot down, and rescued. She never quit, she came back again and again, year after year. She gave hope, love, entertainment, encouragement, and fought to offset Hanoi Hanna's propaganda messages. Chris gave meaning to going home after what may have been the worst day in one's life. She encouraged many to not give up. Thousands of Vietnam vets fondly remember hearing her "Hi, Love" and "Bye, Love" as she started and ended her nightly radio show on Armed Forces Radio.

DONUT DOLLIES

Much of the same can be said of the young women that came to Vietnam to serve with the USO as Donut Dollies. What wonderful girls they were. There to listen and talk, play games, help with letter writing. We would take them into forward areas and let them spend time with the soldiers in the field. After a couple of hours, we would return them to their home station.

HANOI HANNAH (TRINH THI NGO)

I would listen to Hannah as she spoke. She had such a beautiful voice. I would listen to her propaganda messages and think of home.

January 18, 1967 (Wed)

UH-1D
64-13885
Lnds 16
Resupply/support Lane
AHP 26th ROK REG
4:00

We were three aircraft and we approached a mountain peak to extract about 30 refugees in marginal weather. We were only able to put one skid on the top of the mountain while they all got on the aircraft. I pulled pitch to clear the mountain and dove the aircraft down the side of the mountain until I gained enough airspeed to fly. We all three had to make like an airplane and do a running landing. We developed an oil leak and had to ground the aircraft. I suppose we overstressed the aircraft. This afternoon we hauled a light Col. around to Bong Song and a couple of outposts. We flew low level back. We hit some communications wire at the end of a new runway. The wire hooked on the skid and when we landed, we practically landed in a minefield. We took the wire off and flew the Col. back to Bong Song. Note: Bong Song has lost a total of 10 aircraft in the past two weeks.

January 19, 1967

UH-1D
64-13829
Lnds 15
C&C Lane
AHP 1st Div
2:25

INTO THE STORM

Today we flew command and control. We hauled a couple of Col.'s around. 1LT Sid Richardson wrapped a Huey on the side of a mountain. Everyone escaped with no physical injuries.

January 20, 1967 7:45 minutes 50 plus 2 landings

UH-1D
64-13705
Lnds 50
Lane
AHP 2nd & 3rd BN
7:00

UH-1D
64-13705
Lnds 2
Qui Nhon to Tuy Hoa
1st Div
:45

Flew 50 Sorties Resupply to forward camps, all incident free. Even in the mountains to the artillery fire support bases. The Koreans are so disciplined. All day I see a group, surely a different group practicing judo, karate, or something as we fly into, out of, or past. A lot of our flying with the Koreans has been safe. I probably will regret that thought.

We landed in Tuy Hoa tonight; something big is coming up.

January 21, 1967

UH-1D
64-13705
Lnds 31

Combat Assaults
Tuy Hoa
8:30

Oh, what great news in the briefing this morning. Wow. Today we're lifting 1842 Korean troops into the mountains. We were in flights with the 1st Cav as lead. Today was a bad day. The Cav lead aircraft led us into an IFR box canyon while still in formation. Lost sight of the aircraft in which we were in tight formation. Fearing that we would mid-air, I lowered the collective, pushed in the pedal to make us fall very quickly to get to the ground and get out of the canyon. As we began to get close to sea level, I began to pull pitch back in and trim the aircraft back up. I assumed everyone was ok with what I was doing since no one was commenting on the severity of the maneuvers. We broke out with about 100 to 150 feet above the ground and could faintly see some of the slick and gun aircraft flying to get back out of the canyon. Now knowing what we were in for, the AC formulated a plan to try to complete the mission. The clouds went from the deck up. When it lifted (it only lifted a couple of hundred feet), we tried to move the troops. We must've gone IFR 40% of the time. Near misses and accidents was our worst enemy today. After we flew all day, 8 hours and 30 minutes, we landed only to find out transmission gears had practically disintegrated. Call us "Lucky," the maintenance Officer said. He didn't understand why it didn't fail. Gears all ground up in the bottom of the tranny.

I understand that we make incredibly questionable flying decisions here. We do things that we would never do in "real-life". But how does flying 72 helicopters in close formation into a cloud make any sense? I just don't get it. What a way to start the day. Then again, I haven't been here but 12 whole days, maybe that's the way to fly here. When the weather got better, and everything improved, we were just flying with our nose in the dirt. Huge improvement, huge.

INTO THE STORM

January 22, 1967 — no entries in log book.

Well, we left 64-13705 in Tuy Hoa. It needs a new transmission. We hitched a ride back to Qui Nhon and here I sit doing nothing. The weather outside is just how I feel "miserable". Monsoon, the clouds, and rain, we are inside the rainstorm. The buildings we're in remind you of a chicken house. It has a wire screen about six feet up from the ground. Supposed to help keep out the mosquitoes and keep a draft to help cool us down. Any idea how it worked? Class anyone, anyone. Well, what the heck, I went for a walk in the storm. Got a bath and my clothes washed at the same time. How convenient is that. I do believe that if I tilted my head back I would literally have drowned. There is that much water falling. Think I'll go back into the building now. I think I'm losing my marbles or misplaced a few along the way.

January 23, 1967

UH-1D
64-13831
Lnds 6
Qui Nhon
4 sorties
1:15

Well, the hours seem to be passing faster. I hit a bamboo tree low level at 100 knots. Had bamboo splinters jammed into every orifice of the M60 machinegun. We had to clean it. You could see signs on the rotor system and on the horizontal stabilizer. I was lucky that it went down the right side and not the left side. If the left side, then it would have hit the tail rotor. Then we would have had an opportunity to excel… by demonstrating our emergency procedure proficiency or the lack thereof.

In the beginning, one of my letters to Joyce — Jan 67. I know it's sappy, but life is what it is and you're viewing a glimpse into the most intimate part of mine and Joyce's lives. There is a reason why I share all this. That reason is that I'm not the only one that went to Vietnam. Joyce went too, in letters and upon my return the grief, pain, and sorrow that she suffered at my hand. She bares emotional scars that only a few will ever know. Consider as you read or go back and read this that it isn't just me — I am transferring ALL these problems, issues, pain to Joyce. Not proud of it, however, it is just the way it was and continues to be. There's a "RAMBO" spirit in every woman and Joyce is my RAMBO. Never quit. Never stop. Never give-up!

January 24, 1967

> UH-1D
> 65-9575
> Lnds 48
> Sorties 48
> 7:25

In the morning, we were involved in an awful fight. Seems that one of the Korean infantry companies we were assigned to encountered NVA regulars. The small arms fire was terrible with tracer fire easily seen among the smoke and overcast sky. We had been re-supplying them with ammunition and water. We were called in to medically evacuate a Korean soldier to a hospital. He had been shot in the head and in the chest. We don't know if he lived or died. Sucking chest wound. Today we airlifted a four-star general (Westmoreland or Abrams?) around. We saw a lot of contraband today. A large arms cache and drugs. I have got to get some sleep. I don't remember if I ate today or not.

January 25, 1967

> UH-1D
> 64-13833
> Lnds 43
> Lcl Medevac
> 5:30

Koreans' got into the stuff today. Fighting was very heavy and they had quite a few casualties. We resupplied and flew Medevac all day. Got a few bullet holes in the airframe. No one injured on board the aircraft. No sweat. We're all starting to get drunk in order to sleep. Too many sounds and smells to keep you awake.

January 26, 1967

No flying! I went into Qui Nhon and saw the town. Girls selling themselves, families selling their daughters, and wives. Noted major cultural differences and children peeing in the streets. I had my ears cleaned and some oriental chiropractic work done. Bought a camera and a new holster. We were issued Colt .38's and wore holsters that made us look like Wyatt Earp. The old holster rotted. I flew 48 hours and 5 minutes with the 161st Avn Co.

Training is over with them—back to the 4th Infantry Division.

January 27, 1967

Well, today we came back to Pleiku. Spent the day setting up areas and working on the Officers Club. A bamboo building with a thatched roof, bar, and stage. We hired some Montagnards to do most of the work. When I arrived, they were playing a version of mumbly-peg with crossbows. We used to play a game when we were young, where we would take our jack-knives and throw them at the ground and get the opponent to get them to do the splits. Well, we would ultimately get to a point to where we couldn't stretch any further and fell down. Well, these two Montagnard's were doing it with crossbows and apparently, the goal was just how close they could get the bolt to the other opponent. That was most impressive.

January 28, 1967

Went into Pleiku today and stopped by Camp Holloway to get liquor for the Officers' Club. We're still a little disorganized yet. We still don't have our aircraft. They are supposed to be down near Saigon in a bay town by the name of Vung Tau.

January 29, 1967

Nothing happening, we're getting settled back into the Dragon mountain base camp—listening to music. Needed a haircut—we're cutting each other's hair.

January 30, 1967

We've got some slot machines coming. Word is they are trying to hire an Australian stripper to come in and strip one night. Ok, that'll be the day and just what we need.

January 31, 1967—no entries

One of the letters from Joyce in the beginning. Remember that she and I married six weeks before I shipped out.

February 1, 1967

Maybe I've found an alternate career. I'm cutting enlisted guys hair today. We should be going to Vung Tau tomorrow to pick up our aircraft. We've been told that the aircraft arrived weeks ago and have been off loaded and put back together. They mothballed the aircraft when they put them on board a baby aircraft carrier to bring them over. Looking forward to doing something other than dying of boredom.

INTO THE STORM

February 2, 1967

Well, we made it to New Pleiku, but our plane didn't come for us. So, we came back to Dragon Mountain. What! Now we can't get a ride? That probably means that all aircraft were on a mission or down for maintenance. Not necessarily a good thing.

February 3, 1967

Well, today we made it to Saigon. We're just spending the night. You can still see a lot of the French buildings in the more populated parts of town. Played with a French toilet. It had a big wooden box near the ceiling and a long pull chain coming from the box. This is what happens to you when you're in the field too long. Went out and saw the town.

February 4, 1967

Well, we finally made it to Vung Tau. Went to the South China Sea, got on the beach. The waves looked to be about 10 to 20 feet tall. So, I laid down on a towel. In a few minutes, a young lady approached me with a pineapple and a machete. Oh-oh, now what have I got myself into. Well, she's selling them. I purchased one and she peeled it with the machete with precision. I smiled and thanked her. I asked her who were all the people on the beach. She replied people, soldiers, and families. I asked if the soldiers were south Vietnamese? She replied South Vietnamese and VC they both do R&R here. I asked her should I be concerned about the VC, she said no, they needed the rest. I laughed and laid back down. I can't believe I fell asleep on the beach and got a sunburn. I've got 2nd degree burns on my face, chest, legs, feet, and arms. I deserved the sunburn; what idiot falls asleep in the sun. Boy, I am in so much pain and miserable. Oh, I'm going to pay for this for weeks. I can't say anything or if I can't fly I'll get an Article 15. That would be disciplinary action for damaging government property—namely me. Suck it up buttercup—let's see what you are made of now.

February 5, 1967

UH-1D
65-12895
Lnds 3
4:45

Got to get into the flight suit, ouch. Well, today we head back to Pleiku. We got our own aircraft. We flew through Saigon, Phan Thiet, Phan Rang, Cam Ranh Bay, Tuy Hoa, Qui Nhon, An Khe, back into Dragon Mountain. The countryside was beautiful. If there weren't a war going on, it would be a wonderful place to live. We went through the Ia Drang Valley, a free fire zone. We didn't need our attorney to return fire in this area. That was something else discussed on one of our briefings. We could not return fire unless we had clearance from command. If we did and they were "friendlies" we could be court-martialed. Give me a freaking break, would you. So, I emptied a clip from my M-16 through the window in the Ia Drang valley in frustration. Nope didn't help relieve anything.

February 6, 1967

Just loafed, nothing to do. Can you believe it? Listening to music on a reel to reel that I bought from CW2 Bernard Paez. $350, Balance owing $150. (Note: Does he owe me $200 with interest since 1967? He got everything back when I was evacuated).

February 7, 1967

Just read books all day, talk is going around about flying tomorrow.

February 8, 1967

UH-1D

INTO THE STORM

65-12893
Lnds 7
C&C
3:00

I am resting the rest of the day. Sunburn is peeling, man its itchy. My forehead came off in my helmet. A solid piece of skin. I was lucky. I could have been the first officer to get an Article 15 for a sunburn.

UH-1D
65-12893
Lnds 37
Resupply LCL
Dragon Mtn
10:40

Whew, talk about being tired. The governor failed, and we almost crashed into a mountain. We managed to get it back with 6000 RPM. Had to be ultra-smooth with it. Some of the pilot ego stuff one must maintain to do some of this nutty stuff. She didn't want to fly at all. Just any wild movement would make her settle back towards the ground. It takes at least 6000 to do it right and operating is about 6600 RPM. Flew re-supply all day. One LZ was in the middle of a river. Talk about getting wet and no cover. Sometimes I hate being a bird on a string. We took fire all day. Sid let me do most of the flying—thank God. It's terrifying when all you are doing is sitting in the front seat watching all of the tracer rounds fly past you or hear them striking the airframe. Flew with LT Richardson. What is it with all the smoke now—is the whole country on fire? Almost 14 hours today—too long—too long.

February 10, 1967

We are working on our living quarters. Notice the cool way Dick Morrison is standing there with his wild west holster and his Colt .38. The gun would rust in a day. Excuse me, the "weapon" would rust in a day. We are engaged in building top quality furniture in these pictures. We all went for this wild west look.

February 11, 1967

We're desperately trying to create a livable area.

February 13, 1967

UH-1D
65-12895
Lnds 9
CA
3 Tango
2:30

We hadn't flown so long (five whole days) that whoever "they" is decided that we needed some pattern time. We were flying in the pattern when we got called out on a quick mission. Where the heck is it safe here? We heard on the radio that an Australian stripper was at our newly finished Officer's club. When we got back into camp, we bee lined it to the club, we were all having a great time right up to the time the floor caved in. Part of the building collapsed. It was just hilarious. She was so mad and unhappy.

- 1 Combat Assault
- 42 minutes Pattern

I wish they would quit prefacing every mission with "It might be a hot LZ." We understand that already. We are as prepared as we can be and assume all landing zones are going to be hot. Finally, we're getting the word that some General is going to be putting us to work.

February 14, 1967

UH-1D
65-12895
Lnds 37
9 DCS 2/8 Infantry
3 Tango
8:24
9 Direct Combat Assaults

We took fire on approach, drop-off, and take-off. Things are hot today. Seems everywhere you go there are bullets in the air. Tracer fire, little green bugs flying all over the place. I think they're only found in this part of the world. We're flying Red Ball today. You know Ash and Trash (beans, bullets, water, cigs, beer, cokes, ammunition, fragmentation grenades, White Phosphorous grenades, and hot chow for the last flight). We've even got some mail for the troops. There are days when one wishes they are truly indestructible. I was told that we're physical, emotional, and spiritual beings. That being the case, all three took a beating today. Emotionally and physically I'm worn out. Spiritually, let's assume the VC or NVA are loading 1 IN 5 instead of 1 TO 5. That means I can see 1 out of every 5 bullets coming in my direction. Or in the case of 1 TO 5, I can see 1 out of every 6 bullets coming in my direction. That would mean that if I saw about 10 tracer rounds, I had 50 to 60 bullets coming at the aircraft. Now we are behind a paper-thin skin and a thin piece of Plexiglas windshield. How did all those bullets manage to miss us today? Spiritually I'm feeling good since no one got hurt aboard the aircraft today considering that the actual number of rounds expended at us was closer to 500 rounds. Man, I must be a sick dude thinking about this stuff.

When I was publishing some on Facebook, Steven Hyde made a comment regarding the importance of "Ash and Trash". This was my response.

Steven and I know that it was a serious thing to mess with supplies for the troops. The Company/Battalion cases were coveted, they contained cartons of cigarettes, toothpaste, candy, C-rations, P-38's and P-something else and big items. You might ask how I knew. Because I saw someone open one and begin stashing and handing out stuff. When he offered some to me. We exploded and said that stuff is for the troops in the field, not you "cake-eating" SOB's. You want this stuff, go get your canteen, M-16, and we'll take your butt somewhere where you can earn this stuff. He collected the stuff and repacked the box. I think we were all about to go ballistic on him. No one takes anything off the "REDBALL" express. "Ash and Trash", water, ammo, and other supplies, Hot Chow if available, beer, candy, C-rations, maybe a bottle of scotch, cigarettes, toilet paper, etc. Load it ALL till we are FULL. You guys deserved everything and anything we could get for you. Steven, crew members kept us alive many times. You're still my hero's. You guys heroism was almost a daily thing. "When the extraordinary becomes seen as ordinary, it is a sad thing indeed, Rawlinson"

February 15, 1967

UH-1D
65-12895
Lnds 45
12 CA/ 9DCS
2/8 3T
10:30

INTO THE STORM

Supporting: 1st Battalion 22nd Infantry 4th Infantry; 2nd Battalion 8th Infantry 4th Infantry
Location: 3-Tango (Plei Djereng) and LZ501

February 2014 will soon mark 47 years since I was in the region of South Vietnam known as the Central Highlands. It seems so very long ago; however, I still retain clarity of mind over so many minute details. Many of those details are quite humorous and others not so much.

Most helicopter war stories begin with "This is no s**t, there I was," However, this story began for me on February 15, 1967. On the morning of the 15th, we received our briefing at the operations tent. It was a simple mission which involved 72 lift aircraft (UH-1 (B and D-versions)). We were to lift troops from 3-Tango (Plei Djereng) and insert them into a "cold" LZ near the Cambodian border. We were to insert them into an LZ at map coordinates, YA602540, just west of the river. There would be a Command and Control (C&C) aircraft circling to provide aircraft spacing, approach, and departure. The C&C aircraft circling high overhead would be able to see and coordinate with ground forces and aviation assets as required. In the event of fighter and bomber requirements, the C&C aircraft would control the separation of that traffic as well. If the landing zone was "hot", then it became the responsibility of the C&C and the individual pilots and crews to help maintain separation to prevent mid-air collisions. Light observation helicopters such as the OH-23G Raven would perform missions required by the ground and air commander. The weather briefing was unremarkable except for it being hot and getting hotter, with some intermittent clouds, smoke, and haze. Winds were mild and out of the East.

The crew consisted of the aircraft commander, pilot (Donald Rawlinson), crew chief, and gunner. Aircraft number was 65-12895. Once the pre-flight was complete, I pulled the trigger. The igniters fired and ignited the jet fuel; the compressor section of the turbine began its low moan spinning up to its normal high pitch whine. Once at operating

RPM and permission for take-off given, it was up, up and away to 3-Tango. Hurry up and wait was the name of the game. We shot our approach to the refueling area and topped off our fuel load and hovered to the loading area and shut the aircraft down.

Once all aircraft were present and briefings completed, the loading of troops began. Typically, there were between four or five troops, depending on the temperature and humidity. At this point, I was asked what temperature and humidity had to do with anything. This was the simple and short explanation. As the air heats up, the air molecules thin out causing the rotor system and the jet turbine to be less efficient. Since the gas turbines efficiency was reduced, more horsepower was required to perform any heavy lifting. This meant that more pitch had to be put into the blades (rotor system) to lift the aircraft. More pitch in the rotor system meant more torque (power produced by the engine). The heated air combined with high humidity also affected the flight maneuverability or flight performance of the aircraft. Thus, heat and humidity affected the aircraft. The dynamics that could affect power could change in a relatively short period of time within the day. As a rule of thumb, due to cooler temperatures and lower humidity, we could get five fully loaded soldiers on the aircraft in the morning. Since we could produce less lift by mid-morning, we were limited to four fully loaded soldiers due to heat and humidity and maybe some small supplies.

Due to the number of aircraft flying in single file formation, we would form what was referred to as the "long green line." The irony in this was that we formed a long arrow in the sky pointing to our destination.

At this time, I uttered a small prayer for God to get us through the day. We picked up to a hover, checked the instruments (all green), pulled a power check (good enough) and took off to join up with our comrades. We climbed to 1,500 feet above ground level (AGL) to get out of effective small arms range. We were flying at 80 knots (92MPH) at 1,500 feet above the ground or just outside, (or so they say) the effective range of small arms fire. Everything was going well; we were the 3rd aircraft back in single file formation. The lead aircraft reported that the flight was on downwind. The aircraft that were in position began to turn on to the base leg; others followed as the lead aircraft rolled out on final. We were timing our touchdowns so that each aircraft would have about five seconds to offload troops, pull the collective pitch. We were now behind Aircraft 1 and 2 on final approach. As we continued to descend on final approach, I noticed one of the OH-23's sitting out nearly in the middle of the LZ (landing zone), the engine running, and the rotor system was turning and beating the OH-23 to death. It was obvious that no one was around. We saw the tracer fire everywhere, knowing that the ratio of tracer to ball ammunition was 1 to 5 rounds.

I was flying and I looked at the other pilot who said, "Where did the other two aircraft go?" We could see the orange tracer fire coming from the tree line of the friendlies. We also could see the tracer fire coming from the opposite side of the LZ. I said, "I don't know; they just disappeared." I instinctively pulled collective till the RPM started to bleed off. The two-aircraft had either crashed or went down in other locations. As the aircraft began to climb and gain airspeed, I was thinking, "Is anyone behind us or are we the only ones who didn't land (doing a go around) and what do we do now?" We climbed back out to the coordinates of the RP (reporting point) and turned onto the downwind leg. It was at that point we could see that the entire battalion had followed us. We called downwind, then base and final approach. As we descended below the tree line, we could see gunfire coming from both tree lines. We could also see that the enemy had several bunkers they were manning. Plexiglas and thin aluminum offer no protection from even the smallest of offensive weapons. Therefore, we got small in that armor-plated seat.

As we touched down on the heels of the skids, the Crew chief was getting everyone out on the side away from the enemy gunfire. We counted to five, pulled pitch, rocked forward on the tips of the skids, climbed, and gained airspeed as fast as possible. Now we knew what to expect.

During our time away to pick up the troops, Delta Troop (1st of the 10th CAV) was providing on site gun support. Gambler guns (B Company 4th Aviation Battalion) may also have provided support that day. The gunships were outfitted with four equivalent M60 machine guns on a flexible mount. The guns were capable of firing about 550 to 750 rounds per barrel or about 2100 to 3000 rounds a minute. I don't believe they could carry much ammunition except 5,000 to 10,000 rounds per lift. The ground units and the artillery observers (AO) were putting calls out on the "guard frequency 243.0 UHF or 121.5 VHF" for gunship and fighter-bomber support. Normally there were Navy, Marine, Air Force jets and A1E Sky Raiders (Sandie's) on standby somewhere in the country.

As we returned, we could see this nightmare in front of us. The sky had aircraft from 10,000 feet down to the tree level. Napalm (jelled gasoline) and HE (high explosives) were being rained down on two sides of the LZ. There was now a narrow corridor for us to descend through to make our approach and ascend for takeoff. One way in and one way out was/is the ultimate sin in combat helicopter flying. The gunner and crew chief would fire their M-60's until as long as they had the enemy in sight. Once on approach, they could not fire unless they could identify the enemy. This time we dropped the troops and picked up the wounded. One of the troops was wounded before he ever got out of the aircraft. As he was exiting the cargo bay behind me, he was shot by enemy fire. Out we went again: the noise, the stink, the fire, the rounds hitting the fuselage and rotor blades all added up to a very bad day.

We were now carrying wounded and we increased our speed to near VNE (velocity not to exceed) 120 knots was the equivalent of 138 MPH. It could be pushed a little more, the record was about 150. As we approached 3-Tango, we called on the radio and told them to clear us to the aid station because we had wounded. We shot the approach as close

INTO THE STORM

to the aid station as we could without creating problems for them. Once unloaded, we called for fuel, picked up to a 50-foot hover, moved to the refueling area, and sat down next to the fuel blivets (big rubber tubes containing JP-4 or jet fuel). The gunner typically refueled the aircraft while the crew chief was checking the aircraft. While the aircraft was being fueled, this was the only opportunity we had to relieve ourselves. Once those critical tasks were accomplished, we strapped back into the seats and radioed for hovering instructions to the loading area. At that point, we picked up our next load of troops and so began the process over again. We also added some ammunition, water, and grenades.

The process didn't stop until all the wounded troops were evacuated, troops resupplied, or the LZ was closed. We would transfer control of the aircraft (cockpit) workload to help reduce the fatigue and monotony. "You've got the aircraft. I need to reload and rest a few." "Roger, I have the aircraft."

Once all the troops were airlifted into LZ501, we focused on resupplying in bulk. We resupplied, focusing on water, medical supplies, ammunition, grenades, rockets, detonation cord, Bangalore torpedoes, white phosphorous grenades, cigarettes, beer, and cokes. We always tried to ensure that hot chow reached the troops with our last flight of the day.

Some ask, "Weren't you all scared?" Yes, there was a certain amount of fear experienced during the action. However, what was anyone going to do? You can't quit, you can't go home to mother, you have to suck it up and overcome your fear during the battle. Everyone that I flew with that day, all the aviators and crewmembers in that flight, performed amazingly in spite of truly overwhelming conditions. Once the battle was over, everyone dealt with it in his own way.

It was then the crew chief and gunner would assess the true damage, if any, to the aircraft, make repairs, and had it flyable for the next day mission. There was many a night when crew chief and gunner worked all night repairing, servicing, and preparing the aircraft for the next day's flight. Sometimes I would look at the aircraft in awe that someone didn't get shot while we were flying. Positive or negative, I would look back and search my mind, thinking back over the day and question, "Did I cause anyone to die, could I have done something different, something better?"

Aircraft losses that day, 1—OH-23 Light Observation Helicopter, (LOH) 2—UH-1D Helicopters

February 16, 1967

UH-1D
65-12895
Lnds 10
8 CA
Ace High-NPL-Ace High
2:30

We're back at the same place we were yesterday. Seems that we're bringing in replacements for those that died and were wounded. Somber day today.

February 17, 1967

UH-1D
66-748
Lnds 42
2 CA
3 Tango
10:55

Today was a little overwhelming. Too much happening, dying, the smell is more than I can bear. The napalm and burning bodies have a sickening smell; can't seem to get the smell out of my head. The gun fire was so intense at times that we could hear the rounds coming through the cargo area. They were trying to hit the troops even before they could get off the helicopter. We were performing medevac for the troops we were bringing in to the LZ (landing zone) even before they could dismount the aircraft. We were in trouble and we knew it. I fully expected to either crash or die any minute all through the day. Hour after hour they were getting us on pure attrition. Knocking the helicopters down and they were going down everywhere. Or it seemed like it. We were hot refueling at Plei Dejereng (3 Tango) and loading troops. We would jump out of the aircraft and relieve ourselves and jump back in and be ready to go. We never shut down. The crew chief would jump out wash out the blood and if there was time count the bullet holes and give a quick report. To my dismay, he would just shake his head and get back into the aircraft. Back into the air, we would go again. Again, the AC called for air support. The gunships simply weren't carrying enough firepower and what they did carry wasn't powerful enough to do the job. We transmitted on UHF, VHF, and FM and finally started getting some help. I remember the A-1E's coming in—they looked great. The napalm had a sickening smell and was extremely hot. The cannons when they fired we could see the tracers going under the skids at times. It was so intense that I couldn't tell if we were getting ahead or falling behind. Then the Navy fighters came in; they were dropping high-explosives (HE) and Napalm. They would begin their dive at what looked like 10,000 feet and dive straight toward the jungle. At what looked like the last second, they would pull the stick back level out and drop the load and they would shoot straight back into the sky. I envied them. It wasn't until in the afternoon that I could tell we were making progress. It was getting dark, we were bringing in anything and everything that we could: beans, bullets, grenades, claymores, cigarettes, water, C-rats, and beer. We called the last flight the Red Ball Express. When the dark came, we were done; we couldn't go in after dark. I saw an article in Stars and Stripes after the battle. I thought it referenced the battle as Andrew Jackson, however, my wife says that the operation was Sam

Houston. I mailed the article to my dad, he didn't keep it. Our losses appeared to be quite heavy, however, the numbers were never shared. How will I sleep tonight? I've thrown up so much today that I hurt everywhere. Worn out aircraft—jet turbines are failing. Someone said that Lycoming was on strike back in the World.

February 18, 1967

UH-1D
66-747
Lnds 2
1DCS
3 Tango
:30

Whoops—one trip in and Swiss cheese. Nothing like tracers at night to light your way, it's the only way to see the country. Don, kill the navigation lights, they can see us. Oh, by the way how come some of these are orange, aren't they supposed to use green? Suspect they're using captured weapons and ammunition. Trying to determine which way tracers are going is hard sometimes, especially if you're flying. Unless they're coming directly at you, then they look like basketballs.

- 1 Direct Combat Support with 2 landings

February 19, 1967

UH-1D
66-748
Lnds 6
2 DCS
3Tango/Oasis

1:30
LRRP support

How do you land with no lights and the only light is the moon? I hate to put these guys out in bomb craters. We fly them out 20 to 30 miles into nowhere and just leave them. How do they survive? Sometimes they seem happy and other times subdued. It just seems wrong.

February 20, 1967

No fly day. Listening to music and reading.

February 21, 1967

UH-1D
66-748
Lnds 4
1 DCS
Local
1:00

We inserted a LRRP force into nowhere. I did a slope landing (one skid). I landed in a huge bomb crater. I was able to put the entire helicopter down into the hole. I kept just the rotor head above the ground line. Someone said the Long-Range Reconnaissance Patrol (LRRP) was to stay out for six weeks.

February 22, 1967

UH-1D
65-12895
Lnds 22
6 DCS

Local
5:15
Refuel in Oasis and 3Tango

What does one do when you are all alone tens of miles from anyone friendly and there's no one to call? Can't write too much about today. All I can say is there are big fires in Cambodia.

February 23, 1967 — no entries

February 24, 1967
I pulled officer of the guard duty tonight. Now, this was an incredible experience for me. In my mind finally, I'm on the ground with REAL soldiers. Fearless and ready, almost begging the enemy to come closer. Yea right, that was OUR desire alright. We don't want to wake Charlie tonight. I would like to see Charlie sleep all night long. As I stared across "no-mans-land", the area with rows upon rows of barbed wire, claymore mines, foo gas in 55-gallon. barrels. I thought let's not let anything get out of hand. I'm the guy with only eight weeks of basic training. What the heck do I know? Thankfully, a sergeant was there to make sure that the foolish Warrant Officer didn't get himself or the men killed. I was told there are 465 bunkers around the division base camp. I had 25, with one being the command bunker and 24 other sandbag bunkers. The Sergeant showed me where I was supposed to sit. As I sat down, I noticed about 10 klackers, all with wires running to them. As I was reaching and speaking, asking what these things are, he cried out don't touch them yet. Ok, I almost snapped to attention. He said they are detonators and they are attached to explosives, claymores, and jellied gasoline. I said OK, who is responsible for them and who is going to decide if something needs blowing up. His reply, "You are." How will I know when? He calmly tells me that he will tell me when to do anything and everything when the time comes. At this point I'm thinking, "Yes

INTO THE STORM

Sir, Yes Sir, Three Bags Full. You do realize that you can't leave my side now until I'm relieved." He goes, "Oh the guys seem a little nervous. So, you need to call for an infrared (IR) searchlight on a jeep." We need what? I didn't even know we had such a thing. I'm thinking now that sounds cool. When it arrived, the NCO driving the jeep hands me the IR binoculars, so I can scan for movement in and around the perimeter. It appeared to be daylight wherever the searchlight pointed. I couldn't see any movement, but then what did I think I would see. Some guy out in the field waving a flag at me. I handed the "binocs" to my NCO and in the deepest soldier voice, I could muster. I said, "Sarge, you take a look and see what you think." He scans arounds tells the other NCO some cardinal directions and then says, "It's clear." I'm feeling a little out of my comfort zone. I ask "MY" Sergeant if he would show me this other thing that's sitting on the sandbags. We went back to the command bunker and began watching through a starlight scope. He explains that it is a monocular or a single viewing device that captured the available starlight and moonlight and concentrated it so that you could see. It worked well. Cool, do you think it would be alright to show the guys and let them look around? Now I'm sure at this point in the dark he just rolled his eyes. So, I was walking between two of the bunkers when one of the guys tackled me and pulled me down to the ground just as a grenade exploded. I took a small piece of the fragment in the right wrist. Nope, they don't give Purple Hearts for friendly fire. Turned out to be one of two guys that I caught hot-rodding a ½ ton vehicle through the mud. They decided they would frag me for making them wash the vehicle. I was glad that the fella that tackled me had good hearing. Said he heard the striker on the grenade hit the cap. Lucky me. I learned quickly you don't do any discipline like that in a combat zone. Sergeant put a bandage on it and said that it was ok. He cautioned me that some things are just better left alone. Hey, that's all you must tell me.

Well someone got spooked and suddenly one bunker opened fired which caused a domino effect. All 24 of my bunkers lit up the sky with tracers. That caused some of the other bunkers to go ballistic. I was so proud of my Sergeant. He said I'll be right back, and he went outside and talked to the guys and they quit shooting. There was a lot of screaming at the guys and arm waving. After a while, he came back in and told me to take a nap. Ok, my education was growing by leaps and bounds. I think I'll look forward to doing this again.

February 25, 1967

It's a music and reading day.

February 26, 1967

It's a music and reading day.

February 27, 1967

UH-1D
65-12895
Lnds 4
1 DCS
Lcl
:45

We inserted a LRRP force at night into a moonlit bomb crater. Tough to land with no lights, unable to see anything. Used the moon and shadows to land.

February 28, 1967

UH-1D
65-12895
Lnds 9

3 DCS
2:00

Called out for a couple of quick support missions

March 1, 1967

UH-1D
65-12895
Lnds 25
1 CA/18 DCS
Lcl
:30N
8:05

My dadgum attitude is getting the best of me at times. Sometimes it is all I can do to just fly—frustration. If we must fly any instruments it is going to be a little tough. They don't seem to have much set up for our helicopters. How fast can a Visual flight go Instrument—seconds.

March 2, 1967

UH-1D
65-12895
Lnds 10
2 DCS 2/35th
3:20

You have got to be kidding—why did they give us these stupid Geneva Convention Cards. I just learned today that there is a $75 reward or bounty on each helicopter pilot's head and crew member. They just have to cut off the head and bring it back and they collect cash. Sure don't think much of this.

March 3, 1967

> UH-1D
> 66-748
> Lnds
> 11 4:45

There's nothing to say—I stink, I've got blood on me, I'm starving, and people are dying all around me.

March 4, 1967

> UH-1D
> 65-12885
> Lnds 26
> 12 DCS
> 11:30

It's just another day at the office. My mind is foggy tonight. Sometimes, one just has to step back and get a 30,000-foot view. In the grand scheme of things, does what we do matter? "Yes, it does," focus on the troops, nothing else. We flew support today. Ground Commander had us doing some reconnaissance, we were looking for ground movement. Ground troops began having sporadic contact. During the contact, we were to standby at altitude until we received further orders. As necessary we would continue flying Ash and Trash, resupply on the important items.

INTO THE STORM

March 5, 1967

Having to take time off. I'm over regulation hours. Rest and Relaxation.

March 6, 1967

Sleep/Music/Reading Day

March 7, 1967

Sleep/Music/Reading Day

March 8, 1967

UH-1D
65-12895
Lnds 44
Can't say
10:00

The fighting is everywhere. What the heck is going on? I can see smoke and fires forever. The smell — I can't stand the smell. I had bullets come under the seat today and tear up my flight jacket. The damn things ruined my jacket and my camera.

March 9, 1967

UH-1D
65-12775
Lnds 2
2 DCS
Pk-QN-Pk
2:00

We extracted a LRRP force. They sure looked rough. Most of them had practically nothing on. What hadn't rotted off had been torn and ripped by them and the jungle. Made us excited and feel good to be able to pull them out.

March 10, 1967

UH-1D
65-12893
Lnds 25 LCL
7 DCS
7:25

Who cares, I just want to survive attitude. I sit here writing this and I'm thinking what a stupid statement to write down. Everyone here is trying to survive. It is not all about me. I'm not out in the field living from hand to mouth and trying not to trip a mine or get shot. I've got hot food if I want it and a relatively safe place to come back to each night. Quit whining and suck it up, buttercup. We all made it back safely today.

March 11, 1967

Sleep/Read/Listen

Oh, thank you—I need some rest. I need some sleep. Listened to music and finally fell asleep with the headphones on. I've lost so much weight I think I'm almost in 24 pants. I need to eat more.

March 12, 1967 (This is what happens when you crash one and go back and get another one.)

UH-1D
66-750

Lnds 11
1/22
7 DCS
4:05

UH-1D
65-12889
Lnds 13
1/22
3 CA
1:30

UH-1D
65-12889
Lnds 9
1/22
4 DCS
2:10

We came to a hover over the hole in the jungle and I began to let us down. Suddenly, the aircraft just fell at a very high rate of decent. I couldn't tell if the turbine or short shaft had failed (it just happened too fast). Just before impact, I pulled the pitch, too little too late, we still impacted the ground. As we bounced back into the air I could hear the turbine whine back up. Something screwy had happened and none of us had an explanation for it. I was able to recover the aircraft and bring it to a hover. As I set the aircraft down, she squatted at an unusual angle and slightly cockeyed. I could tell that the skids had flared. Upon inspection of the cabin, we could see evidence of crushed bulkheads. We had hit so hard that the exhaust gas from the turbine had been deflected down towards the ground and had started a fire in the landing zone (LZ). Cpt. Recher and I were flying together today.

We discussed whether we should fly it back. We inspected the aircraft carefully, checked the instruments, and considered our location. We both agreed to fly it back. Shouldn't have, but we did it anyway. We immediately got another aircraft off the flight line and went back out. Thought I broke my back. We all definitely hurt our backs.

March 13, 1967

Dirty, mismatched fatigues, non-subdued patches, dirty boots, at least I didn't need a haircut. A little under nourished or bony. The smile is for you, sweetheart.

March 14, 1967

Sleep/Listen/Read

March 15, 1967

UH-1D
66-751
Lnds 7
Pattern
:30

This was a standards (check ride) ride due to the crashed and busted aircraft on the 12th. What a stupid ride. We initiated an autorotation at 5000 feet with a 3000-foot recovery. What was that supposed to prove? The SIP scared the crap out of me; he obviously had been here too long. We floundered about trying to stay alive at 3,000 feet low rpm audio blearing after bleeding off rotor RPM. Now our own people are trying to kill me. "Yea, Don just pick out an imaginary runway, field, road at 3,000 feet and shoot an autorotation to it. Uh, it's blue sky, you want the throttle at detent or full power, ok I'm going to rotate at 3,100 and pull pitch at 3,010 feet, right? Yes, sure OK. So, he pulls in pitch at 3,010 along with me and now we're pulling RPM down to 5,800. We're lucky to be alive.

March 16, 1967
Sleep/Read/Listen

March 17, 1967

UH-1D
65-12895
Lnds 23 LCL
18 DCS
6:35

UH-1D
65-12895
Lnds 1
6 CA
1:20

There are no easy days — every day is hard at times.

March 18, 1967
Sleep/Read/Listen

March 19, 1967

UH-1D
65-12895
Lnds 25 LCL
6 DCS
6:48

UH-1D
65-12895
Lnds 26LCL
7 CA
2:00

Death is everywhere; you can't trust anyone. Depend on yourself; don't trust anyone.

March 20, 1967

UH-1D
66-747
Lnds 7
LCL
7 DCS
2:35

What did I do to deserve this? This is a screw up—has to be. Sid and I have been given the duty to take Jayne Mansfield (Hollywood actress/model) to outposts, artillery sites, 3 tango (Plei Dejereng), and Oasis. Our criterion for making an approach into an area was basically a radio call, call for pop smoke, and safety check and land. We wanted as many guys to see her as possible.

Flew Jayne Mansfield and her agent (husband). We picked them up at Camp Holloway. We didn't really have any bad times in the areas while we were there. However, everyone could tell that the troops had been in the crap. No one really had to spell it out. I think that really scared her to see just how close she was to the fighting. This evening after dark there were shots fired around the division perimeter. Heavy machine gun fire broke out. She couldn't stand it. It looked like she was going to have a nervous breakdown. We had a no-fly policy that night due to the enemies' close proximity. But in her condition, it was evident that she wasn't going to be able to handle it. The AC decided (after much pleading) to fly her out in the dark any way. I have never seen anyone that scared in my life. Bad news, we had tracer fire tracking us as we took off. I pulled max power, accelerated and pulled a cyclic climb to escape the tracer fire. Sid flew the rest of the way and flew back. I got her autograph on a Military Payment Certificate (MPC). *(See above.)*

March 21, 1967 — no entries

March 22, 1967

    UH-1D
    65-12895
    Lnds 17 LCL
    7 DCS
    2:45

    UH-1D
    65-12895
    Lnds 23 Plei Me Trap
    8 CA
    8:00

BATTLE OF POLEI DOC, MARCH 22, 1967, REPUBLIC OF VIETNAM

On 22 March, Company A and B, 1st battalion, 8th Infantry made contact with an estimated NVA battalion at grid coordinates YA686344
    1SG David McNerney, Medal of Honor
    Aircraft Commander: 1LT Sid Richardson
    Pilot: WO1 Donald Rawlinson

Crew chief: Unknown
Gunner: Albert "Swede" Ekstrom
"One can't guess what today will hold. None are the same; even the boring ones." [Rawlinson]
42 Wounded—22 Killed in Action—A Company
3 Wounded—5 Killed in Action—B Company

Mission briefing was simple enough; we were to be 2nd up aircraft for A & B companies, 1st of the 8th Infantry, also serve in a Command & Control capacity if required. After gathering our personal gear and our Signal Operating Instructions (SOI) books, (Sid Richardson and I hung those around our neck), the SOI contained a record of all the frequencies assigned to all units in Vietnam for that day. Sid and I continued to the flight line to preflight 65-12895 (Huey UH-1D). After performing a short preflight and reviewing the Dash 12, the "bird" was ready for flight; we settled into the seats. I always swiveled my gun belt around so that my .38 was positioned over my crotch. Normally my flak jacket was in the chin bubble in hopes that it would deflect any rounds penetrating in that area. Once cockpit procedures were complete, the aircraft startup and run-up procedures were completed. Once all the cockpit workload was out of the way and the tower called for "hover taxi instructions," we were ready to go to work.

We took off and proceeded to the forward area of A and B Company of 1st of the 8th Infantry. The 1st up aircraft was already out with A Company. We could hear the radio traffic and knew that A Company was engaged with the enemy. The company was on a mission to locate a missing Long-Range Reconnaissance Patrol when they became engaged in a fire fight with soldiers from the North Vietnamese Army. We had landed at the staging area and shutdown the aircraft. We heard the emergency call come from the 1st up aircraft as they broke over the tree tops. They were streaming smoke and looked like they were on fire. They landed quickly and dismounted the aircraft. The crew chief immediately grabbed the fire extinguisher and opened the maintenance hatch for the tail-rotor servo (hydraulics) and put the fire out.

The pilots relayed to us the location (grid coordinates) and what they believed to be happening on the ground. It was only about 0745 in the morning; the day was not starting well at all. With 1st UP now grounded, 2nd UP was now at work. We immediately took off and proceeded to the approximate location of the unit under attack. As we crossed over into the valley, we could only see one spot that had any kind of faint smoke or haze. So, we made our way towards the smoke and came to a hover directly over the trees concealing the activity/fighting below. Since the jungle was made up of triple canopy growth, we could easily be 100-150 plus feet above their location. It was initially impossible to see anyone on the ground.

Sid was on the radio and I was on the controls. Sid was communicating with the 1SG and it sounded bad. Machinegun fire could be heard over the radio and it sounded like it was being directed toward the aircraft. The headquarters had been wiped out. They were under attack by a superior force. We asked them what we could do for them. The 1SG replied that we could provide them with chainsaws, fuel, water, grenades, and ammo. He added that we needed to hurry. We made a mental note of some obvious landmarks and began our journey to 3-Tango for supplies.

Radio Call to 3 Tango: "3 Tango Blackjack 895 requesting the following supplies to be loaded upon arrival: chainsaws, mixed fuel for chainsaws, water, ammunition for M-16, M-60, M-79 grenades, fragmentation grenades, and rope. The unit is experiencing heavy fire and casualties." After a few minutes, we made a second radio call: "3 Tango Blackjack 895 turning final for high speed approach to pick up location." Quick turnaround time was what was needed.

Upon returning to the location, we had difficulty finding the unit. The smoke now was coming up in a totally different location; I lowered the belly of the Huey into the tops of the trees in hopes the friendlies would be able to hear and see us. Enemy fire was directed toward the aircraft and some of the rounds struck the aircraft. Essentially, we asked the 1SG if he could hear us and, if so, were we moving away from him or towards him. Once we located them, I again picked several reference points that would guide us back to their location. "As we all know second time is a charm."

Enemy fire was directed at the aircraft numerous times during the morning and afternoon. Rounds had hit the fuselage, rotor blades, and the first aid kit behind my head. The holes created by the rounds that passed through the main rotor blades had a whistling sound they emitted. Any rounds that penetrated the floor had to be covered to evacuate wounded. The crew chief would normally cover the holes with a poncho, so the wounded were not injured worse on the jagged parts of the floor.

Once the unit was found, the crew began the arduous task of lowering supplies to the ground. This task exposed them numerous times to enemy fire. I would hover the aircraft and then 2LT Richardson would hover for a while. We would divide up the cockpit workload to give one another a break. Dividing the cockpit workload allowed us to reduce the fatigue experienced by pilots during high levels of concentration. Once the key supplies were lowered to the ground via rope, then the dangerous process of clearing a landing zone resumed. Throughout the process of lowering the supplies, we could hear the automatic weapons fire being directed towards the enemy and towards us.

We were also responsible for supporting B Company that day; therefore, contact was required for status and requirements. They were on a track to rendezvous with A company and were engaging the enemy. They experienced casualties in sporadic encounters. As they were closing in on A Company's position, the fighting was escalating as B Company closed the gap on A Company.

We pulled out of A Company's location and went back for additional supplies. Upon returning after 35-40 minutes, we took up a position at 1500 feet above ground level and entered slow flight and worked in a command and control capacity until the Landing Zone (LZ) was completed to an acceptable first entry. It was during this time we were able to see additional firepower unleashed on the valley below. Army gunships (Gambler Guns) were communicating with the 1SG who was coordinating the machine gun and rocket fire. We had Navy jets that were dropping butterfly bombs and napalm which would scatter across the tops of the trees before settling down to the jungle floor below. The high explosive rounds coming in were creating huge concussion waves. Felt like we were getting hit in the chest.

On our first landing into A Company's location, it was very tight. The troops had been using the chainsaws to cut trees; however, at least one chainsaw had been shot out of one of the soldier's hands. The 1SG called for the troops to switch to detonation cord (DET Cord).

(Detonating cord (also called detonation cord, detacord, det. cord, detcord, primer cord or sun cord) is a thin, flexible plastic tube usually filled with pentaerythritol tetranitrate (PETN, pentrite). With the PETN exploding at a rate of approximately 4 miles per second, any common length of detonation cord appears to explode instantaneously.)

It was during this first landing that I had to hover forward and make contact with a tree using the tie-down point on the end of the main rotor blade and strip the bark off while descending into the LZ. The use of DET Cord made some of the tree stumps too tall and I had to move the tail rotor between trees in order to get low enough to unload supplies and load wounded. We saw the panel that 1SG McNerney had placed in the trees. It made it easy to shoot the approach the first time to the top of the trees.

On our first descent, Swede (gunner) was standing outside the aircraft on the skid with his M-60 in his hand. The crew chief and gunner (Swede) hung out of the aircraft while we descended into the LZ. Both gave critical guidance continually as to which way to move the tail rotor, so we wouldn't impact any trees or debris. This process of resupplying continued until an LZ was sufficiently cut in the jungle to allow actual contact with the ground. Once the aircraft was able to get into the LZ completely, we made repeated landings resupplying and extracting wounded. Throughout the day, the smell of body fluids such as urine, feces, and blood all begin to take their toll on us. On one trip into 3-Tango, while the gunner was refueling the aircraft, the crew chief washed out the cargo area. I looked back at what appeared to be a river of blood that came out my side of the cargo area. This helped, but we were already nauseated.

Once while returning and preparing to land in the LZ, another Huey was on the ground. I suspected the crew of the originally assigned aircraft had returned to Dragon Mountain and grabbed another Huey and returned. That was the only time we saw any additional support helicopters. That was a total of three support Huey's during the day. Probably only two crews.

We encountered a news crew that day late in the afternoon. We had little patience for news crews and cameramen. They took up valuable space on the aircraft and reduced the number of supplies that could be airlifted. In the event that a decision had to be made between them, supplies and soldiers, the latter always won out. We would not lift the news crew out until the wounded had been evacuated and the unit resupplied. (Since they stayed, one would think that somewhere there are pictures and possibly interviews.) Sorry, but who comes first, the reporter or the soldier. No question to answer there.

We continued to resupply both A and B company and extract wounded from both A and B company. We resupplied B company with ammo and water on at least two occasions as they made their final push into A Company.

"As days go this was a good day; we didn't die or break anything. God is faithful to provide," Don Rawlinson.

When 1st LT Sid Richardson had the controls, I looked back into the cargo area and out the left side and could see the wounded being lifted into the aircraft. Images seared into your brain that you cannot get out. Due to the severity of the action and the stress, I truly don't believe that our brain can process the information and make sense out of it. So, another day ends. Tomorrow is only a few hours away.

To clear a landing zone for helicopters, McNerney braved heavy enemy machine gun fire to collect demolition equipment abandoned in rucksacks that had been dropped early in the battle, now outside his company's perimeter. Facing continued heavy small arms fire, he returned to the company's location and blew up trees to create a landing site so a helicopter could extract the wounded. Company A fought for almost the entire day. One soldier propped Lt. Sauer against a tree so he could continue shooting. Sauer later said, "By that point, we were fighting for each other. We were fighting to live. Everything (McNerney) taught them, just clicked. They fought for each other. A lot of them did just unbelievable things. They were all heroes that day. No one thought of themselves."[6] Due to the intense action, many casualties were not evacuated until late in the afternoon. Company B, several kilometers from Company A, fought all day to reach the unit, reaching them late in the day. Despite his own injuries, McNerney refused to be evacuated and remained with A Company until a new commander arrived the next day. McNerney said later on that he had a calm feeling and knew that he wasn't going to die that day.[1][7]

During the Battle of Polei Doc, A Company's (totaling 108 troops) casualties included 22 men killed in action and 42 wounded, a 60% casualty rate. After the battle, U.S. troops identified 139 NVA KIA around the A 1/8 perimeter, and another 400 NVA graves were located in the area a short time later.[6] As a result of their actions during the battle that day, two men were awarded the Distinguished Service Cross, seven the Silver Star, 25 the Bronze Star, and 65 a Purple Heart.[6] McNerney was awarded the Medal of Honor. To evacuate wounded and bring in fresh supplies, helicopter pilot CWO Donald Rawlinson repeatedly returned to the company's position despite heavy enemy small-arms fire. Lt. Col. Rick Sauer later said, "He came and flew out the wounded. He risked his life multiple times. ... The back of his Huey helicopter was just flowing in blood from continually taking out the wounded people. He did quite a job."[6] Rawlinson was recognized with the Distinguished Flying Cross and was adopted as a member of A Company by its troops. [7] The battalion was one of the most highly decorated battalions of the Vietnam War. Four members were recognized with Medals of Honor within a 60-day period.

Congressional Medal of Honor (CMH) — presented to McNERNEY, DAVID H.
Rank and organization — First Sergeant, U.S. Army, Company A, 1st Battalion, 8th Infantry, 4th Infantry Division
Place and date — Polei Doc, Republic of Vietnam, 22 March 1967
Entered service at — Fort Bliss, Texas
Born — 2 June 1931, Lowell, Massachusetts
Citation:

1st Sgt. McNerney distinguished himself when his unit was attacked by a North Vietnamese battalion near Polei Doc. Running through the hail of enemy fire to the area of heaviest contact, he was assisting in the development of a defensive perimeter when he encountered several enemy at close range. He killed the enemy but was painfully injured when blown from his feet by a grenade. In spite of this injury, he assaulted and destroyed an enemy machine gun position that had pinned down five of his comrades beyond the defensive line. Upon learning his commander and artillery forward observer had been killed, he assumed command of the company. He adjusted artillery fire to within 20 meters of the position in a daring measure to repulse enemy assaults. When the smoke grenades used to mark the position were gone, he moved into a nearby clearing to designate the location to friendly aircraft. In spite of enemy fire, he remained exposed until he was certain the position was spotted and then climbed into a tree and tied the identification panel to its highest branches. Then he moved among his men readjusting their position, encouraging the defenders, and checking the wounded. As the hostile assaults slackened, he began clearing a helicopter landing site to evacuate the wounded. When explosives were needed to remove large trees, he crawled outside the relative safety of his perimeter to collect demolition material from abandoned rucksacks. Moving through a fusillade of fire he returned with the explosives that were vital to the clearing of the landing zone. Disregarding the pain of his injury and refusing medical evacuation, First Sgt. McNerney remained with his unit until the next day when the new commander arrived. First Sgt. McNerney's outstanding heroism and leadership were inspirational to his comrades. His actions were in keeping with the highest traditions of the U.S. Army and reflect great credit upon himself and the Armed Forces of his country.

Distinguished Flying Cross (DFC) — presented to RAWLINSON, DONALD L.
Rank and organization — WO1, A Company 4th Aviation Bn (Airmobile), 4th Infantry Division
Place and date — Polei Doc, Republic of Vietnam, 22 March 1967
Entered service at — Fort Wayne Detroit, MI/Fort Knox, Kentucky
Born — 27 March 1946, Kalamazoo, Michigan
Distinguished Flying Cross Citation:

Serving as Pilot of an UH-1B helicopter which was re-supplying Companies A and B, 1st Battalion, 8th Infantry who had engaged the North Vietnamese Army near the Cambodian Border. Arriving at the company's positions it was discovered that the landing zone had not been completed and the supplies would have to be lowered to the ground by rope. Although intense enemy sniper fire was directed at the aircraft, WO1 Donald Rawlinson hovered the helicopter at tree top level while the supplies were lowered to the ground. When the landing zone was completed, he made several landings, despite heavy enemy fire, to evacuate the wounded infantryman. Through his untiring efforts all the casualties were evacuated and the companies re-supplied.

A documentary *Honor in the Valley of Tears* described McNerney's service with Company A in Vietnam and his Medal of Honor action. It premiered at the May 2010 GI Film Festival. The film was co-written by the son of one of McNerney's soldiers. Donald Rawlinson appears in the documentary recounting that day's flight activities.

SOI—Signal Operating Instructions contained call signs and frequencies of most military units in Vietnam. It was a small booklet approximately 6 inches square with a neck lanyard on it. It was worn about the neck of the aviators. We all had orders to prevent the enemy from capturing the document.

SOP—Standard Operating Procedure.
UHF—Ultra-High Frequency
VHF—Very High Frequency
FM—Frequency Modulation
HF—High Frequency
Navigation Radios

As one might imagine, negotiating around all the air traffic was troublesome at times. However, all helicopter pilots were trained intensely to clear themselves (look left, right, above and below) for any type of aircraft. We were trained to interpret the surroundings and adapt to the environment. I know this sounds like fluff. However, our instructors were adamant about us being aware of hazards in the air, on the ground, and even in the water. Most were unaware that aviation accidents are the biggest killers in a combat environment. Keeping that in mind, during the day these were some of the potential hazards we encountered:

1. Artillery: We were trained to perform AO duties and call for

fire support. We knew the location of the artillery batteries and the direction of fire. It was pretty easy to avoid. As long as the rounds were accurate and not errantly long, we would be ok. "Faith"

2. Fast Movers: A1E's (affectionately referred to as "Sandies"), and Jets meant we had to stay parallel and within bounds of their runs. Trying to time our getting in and getting out (using a basic corridor) based on their actual firing or bombing runs. Sometimes with the napalm, it would be very hot and pull the air out of our lungs. The A1-E's got pretty close to us and that meant they got close to the troops on the ground.

3. Gunships were pretty easy with six ships in all, three on each side making gun runs. Each equipped with the M6 Armament Subsystem. This system carried four (two each side) M60C 7.62mm machineguns on a flexible mount. Each M60C with a cyclic rate of about 550-750 rounds per minute; therefore, all four could expend 2,200 or up to 3,000 rounds a minute. They carried about 6,800-10,000 rounds on the aircraft. They would fire short bursts multiple times on each successive gun run. You could see where, if they fired continuously, they would exhaust the ammunition load in less than three minutes. We could easily see them and maneuver in and around them. We actually made radio contact with Gambler Guns and asked them to continue their runs dry while we made one run in under fire. We were hoping that the enemy would keep their heads down just for the moment while we landed.

4. It truly was a crowed sky. With artillery, bombs, napalm, machine gun fire, sounds, and smells, it was so noisy at times we could not hear what was being transmitted on the radio. The radio was a fantastic tool and, in this battle, like most, we monitored the following:

- Friendlies frequency
- Our aviation company frequency
- VHF frequency for aircraft and other
- UHF frequency for aircraft engaged in the battle as well as command and control
- So we were monitoring four radio frequencies for traffic, which helped, and the fact that we could transfer control of the aircraft to the other pilot from time to time made the day more bearable.
- "3 Tango, 3 Tango Blackjack 895 requesting straight in approach to the aid station with wounded on board. Over."

5. References:
6. http://en.wikipedia.org/wiki/David_H._McNerney

March 23, 1967

Reading & Listening to Music

March 24, 1967

Reading & Listening — I wrote some poetry today. Seems that many of the guys are writing poetry, I guess that's how we're working out some of our frustration. If I read one more thing about what's going on at home — I'm going to puke.

March 25, 1967

UH-1D
65-12894
Lnds 41
LCL
10 DCS
9:20

LZ regular—this was one assault in which we were sling loading a load of Bangalore torpedoes into a position when we began to receive mortar fire. The aircraft behind us was carrying the blasting caps. We didn't carry the explosives and the electrical detonators on the same aircraft. Helicopters generate tremendous amounts of static electricity. The static discharge makes it possible to accidently detonate the blasting caps. Wouldn't be a good thing if the caps and explosives were on the same aircraft. One of the soldiers began to wave us out of the LZ, they wanted us out of there. We couldn't tell what the problem was until the ground exploded in front of us. I grabbed the stick, hit the pinky button, dropped the explosives, and pulled the pitch. On the way out, our gunner saw the VC forward observer in a tree. He requested permission to fire and I cleared him to fire. Another great day, we all returned safely.

March 26, 1967 (Easter)

UH-1D
65-12893
Lnds 17 LCL
5 DCS
5:15

Doesn't seem right to be doing this today.

March 27, 1967 (Birthday)
Turned 21 today, I'm now considered to be one of the old men. The guys got steaks from somewhere and I wasn't about to ask. Of course, maybe it wasn't cow as they say. Probably water buffalo, however, it could have been monkey. No, it was too big to be monkey.

March 28, 1967

　UH-1D
　66-749
　Lnds 35
　LCL
　10 DCS
　9:30

This will never end, and I will never be the same again. I feel like I'm in my 60's. I've seen too much too young. How will I reconcile this crap when it's over? Well, some people can big man their way on any topic, but if you can reconcile all the things that have happened here I don't understand how you do it.

March 29, 1967
　Reading & Listening

March 30, 1967

　UH-1D
　65-12895
　Lnds 13 LCL
　1 CA
　3:25

We performed a Long-Range Reconnaissance Patrol (LRRP) insertion to somewhere on someone's map. Very close and took fire on the pull out. We had some problems close to the division base camp.

March 31, 1967

> UH-1D 65-12895
> Lnds 14
> 2:48

We flew flare duty tonight. The flares are large metallic canisters. We had one ignite inside the cabin. They got it out before it melted the floor. We dropped flares and took tracer fire for a while. Tracer fire at night when directed towards me always reminded me of big basketballs on fire with big flames. The enemy tracers were supposed to be green. The first time I saw green tracers I had absolutely no clue as to what it was. What does it mean when the tracers fired at you are orange. I always told myself that it was captured ammunition.

April 1, 1967

> UH-1D
> 65-12895 Lnds 14
> 4 DCS
> 3:30
> 1st/22nd Paymaster (1LT Babcock) insertion Montagnard -village
> April Fool's Day Wound — No Joke
> WO1 Donald Rawlinson
> April 1, 1967

What makes a memory? A smell, sight, sound, touch, taste, it can be any of these or a multiple of these. All of us have experiences in life that we will never forget. For some of us, we can only hope that the intensity will fade.

This one incident in my life began not on April 1st, 1967, but on March 31st. The 4th Division Base Camp had some anticipated contact on that night. We were on flare duty. On this night and the aircraft was loaded with large metallic canisters, magnesium flares. The aircraft was ready for a tactical startup and all our equipment was ready for us to jump in and go. With all being ready, the waiting began for the call to come for us to take and begin dropping flares.

The night was warm, humid and sticky. There was smoke in the air; its smell blended with the JP-4 (jet fuel) and the oil that was sprayed on the clay to keep the dust down. I moved to the pilot's seat and sat down just to rest, smells of leather, sweat soaked seat belts, oils, hydraulic fluids permeate the cabin.

I looked around at each one of us and we showed signs of being tired and stressed out. No one was interested in talking; each had taken up a place in the aircraft and was waiting for the call. Once the call came in, as the pilot I would immediately throw the battery switch, pull the trigger to fire the igniters. As the rotor begins to turn it will sling off the tie down strap and gain RPM to idle. At this point, everyone was hooking up to the radios and locking seatbelts and harnesses. That night I'm first to pilot. I began hovering toward the takeoff pad to prepare the aircraft for takeoff. The Aircraft Commander was making the radio calls and receiving the assignment. Anyway, that was the way we planned it. How did it actually go? Well at about 0045 the harassment begins.

Sortie 1 65-12895

Operations sent a jeep to our revetment and advised us to startup and idle, remain on standby. In about 45 minutes the call came over the radio. Roll throttle up, pull the collective pitch, drop nose, pull more collective climbing near vertically to 1500 feet over the Division Base Camp. We immediately began moving to the area on the perimeter that was receiving fire. The gunner pushed one of the canisters and the crew chief pulled the pin on the canister as it went out the aircraft door. Then a brilliant white light appeared off the starboard side and aft of the aircraft. We could see the ribbons of tracer fire from the bunkers dancing and licking the earth as they performed their dance of death. An occasional group of green tracer rounds would come up in our direction. We began a slow racetrack pattern to drop additional flares as needed. The green tracers that were rising towards us appeared to be in slow motion. They were flying by at varying distances from the aircraft. It was not good that I knew that there is one tracer for every five rounds passing by us. We continued to fly and watch the light show below until its conclusion. The sortie was completed in 1 hour and 35 minutes.

With the interaction completed for now, it was time to return for fuel, flares, and await the next mission.

Sortie 2 Aircraft UH-1 Huey 65-12895

On sortie number 2, we were sitting on one of the take-off pads idling. The wind was whipping the warm thick jet exhaust around to the front, making me nauseous. That smell when you're tired, hungry, sweaty, stressed out, and generally just fed up was plain sickening. It was about 0330 when we received our next call. The other pilot was flying this time and I operated the radio and contacted the tower. As we took off; we made a fast-vertical ascent straight to 1500ft AGL and he positioned the aircraft over the bunkers. Out the door goes a magnesium flare and the brilliant light turned night into day. The green tracers and orange tracers exploded in a high tempo dance as they crisscrossed one another. Little green bugs were everywhere. The tracers were hitting objects on the ground and ricocheting into the air and some appeared to bounce from object to object. Some of the tracers appeared to have a mind of their own as they appeared to be hunting for us. The light shows finally came to an end after 1 hour and 15 minutes.

We returned to the revetment (parking area) exhausted, in need of a shower and bed. We lived in such luxury; our GP medium tent housed about 10 of us. I slept on a wonderful army cot covered with a mummy sleeping bag. On the mummy bag were layers of dust and dirt from the countless rounds the artillery shooting over our tent. All the comforts of home. It was almost 0500 when I laid down to sleep.

As one might have guessed, I was in bed and asleep when they came in and woke me. It was 0600, I was explaining to them that I had just laid down and went to sleep. They told me they don't care, that I needed to get down to the flight line. They needed a right seat for a mission with Paez. I dressed quickly and moved to get something to eat. I went to the flight line to find out what was going on. The day was April Fool's Day, so I assumed it was a joke. I was three months into my tour of duty and had accumulated over 200 hours of combat time.

Our aircraft assignment was UH-1D, 65-12895, I commented to myself, "didn't I just get out of this aircraft." Our mission assignment was to fly to the 1st/22nd forward area, pick up 1LT Babcock, XO and a Sergeant (performing Paymaster Duties) and insert them into a Montagnard village.

We took off at 0700 and flew toward the forward area. They were not ready yet, so we shot some touch and go's (practice landings) and finally, we landed for the pickup. Once we had been briefed on location, strength, frequencies, we took off and began our trek toward the Montagnard village. Upon arriving over the troop's location, the Aircraft Commander (AC) called the ground troops for identification and LZ. I circled the area looking at the terrain, grass, wind direction, and speed. The AC requested the ground commander to pop smoke. Then the AC identified the color of the smoke over the radio for confirmation. I continued to circle to line up on final approach. Once on final, I initiated a steep approach into the LZ, shooting to the ground. 1LT Babcock and the SGT dismounted the aircraft out the right side and moved away towards the right rear of the aircraft.

I set up for a maximum performance takeoff. As we began to lift off, I saw three people step out from the wood line.

Would you like to know what helpless really feels like? Hand three people AK-47 fully automatic rifles with 30 round clips while sitting 30 feet in the air behind 5/16" of Plexiglas and tell them to open fire on you.

At this point, we heard the popping sound. It's "funny" how automatic weapons fire sounds just like popcorn. They were on the right side and began firing as we went through about 30-40 feet. The rounds raked the aircraft on the right side from nose to tail. I saw the results of rounds hitting the nose, chin bubble, felt them hitting under my seat, felt them hitting the rotor system, several hit instruments directly in front of me. A couple of the instruments were forced partially out of the instrument panel. I felt the Huey shudder and then my left foot came up to my chest. The blood splattered across the instrument console. The top of my boot had exploded as the round penetrated the anti-torque pedal, the steel shank in my jungle boot, and entered just behind my big toe and traveled diagonally through my foot until it came out on top where the tongue and eyes came together.

As my left foot came up off the left anti-torque pedal my right foot pushed the right pedal, causing the aircraft to yaw violently to the left. I tried to use the intercom to tell the AC I was hit. But since it was the left foot that would depress the floor microphone button, I was unable to find it. He couldn't hear me, so I shouted out that I was hit. It was procedure to turn over control of the aircraft until the extent of the injury was known. I knew I was incapacitated so I gave control of the aircraft to the other pilot. The AC was now piloting the aircraft, crew members consisted of the following: AC, Pilot, Crew Chief and Door Gunner. The Door Gunner was hit by shrapnel in the face and hand and was bleeding profusely. The wind made it look much worse than it was. He had taken a round in the center sternum pocket of the ceramic chest protector, thankfully it did not penetrate his chest protector.

I saw the front of the AC's flak jacket jump and I knew that a round went across his belly or in it. I could see metal tearing and smoke filling the cockpit. The shooting was over in seconds.

It looked like every light on the caution panel was coming on. The engine oil pressure was falling to zero and transmission oil pressure was zero. We had lost hydraulic pressure and the emergency backup pressure in the first seconds of the attack. We had no hydraulics and therefore we had no control over the aircraft. With the loss of the hydraulics, the following few control movements consumed the emergency pressure that was trapped in the irreversible valves. The controls were frozen. The engine oil pressure was now zero and the temperature was rising. The transmission temperature was already in the red. I called out the condition of the aircraft, instruments in the red systems failed or failing, hydraulics, engine, transmission, to the AC who was desperately trying to coax a dead ship to fly.

The aircraft had leveled off about 50 above the trees and was traveling about 80-90 knots. We flew about one to two minutes when the engine exploded. I had made mayday calls on the VHF, UHF, and FM frequencies. I believed that the only radio that was still working was the FM radio. No one responded so I didn't know if anyone heard or not. The aircraft began to settle toward the ground. The AC was attempting to input control movements, however, with no hydraulics and no pressure trapped in the reversible valves the aircraft was uncontrollable. I began to shut everything off, fuel and battery. If possible, I didn't want us to explode or catch fire upon impact. As the rotor system was losing the valuable stored inertia (RPM) in the rotor system, the aircraft was losing altitude. With the loss of RPM, the nose of the aircraft began to yaw to the left. At this point, the right seat which I occupied now had a front row view of the jungle passing by underneath and the ground coming at us.

We broke over an open area and the aircraft continued to lose altitude until we struck a dike. At this point we lost the tail rotor; the aircraft's attitude was nose high and left, then the skids impacted the ground. The aircraft rolled straight forward, and I could see the dirt coming directly at my windshield. As the aircraft pitched forward, the rotor system impacted and dug into the ground. The aircraft snapped up into the air. I could see the rotor blades shatter and the rotor blades break off and fly away from us. As the aircraft crashed back into the ground we came to a rest on the belly of the aircraft.

I got out of the aircraft on the right side with my M-16. I was hopping on my right foot. I looked at the gunner; he was bleeding from the face, neck, and hand. He was wearing his "chicken plate" and had been hit directly in the pocket on the sternum. His M-60 was trashed and the shrapnel that had come off the machine gun had hit him in the face, shoulder, and hands. I looked at the AC through the windshield; he was doing something with the aircraft. I unsnapped the gunner and he got out of the aircraft and walked around the aircraft and met up with the Crew chief. I hopped around the front of the aircraft. At this point I put my foot down and began walking on it, the sound of liquid in my boot and the crunching of breaking bones quickly caught my attention. I looked down and the blood was running out the top of my boot. It was at that point, I could see that the gunner, crew chief, and the AC were walking away together.

I headed in their direction when after only a few steps; I fell. At this point, I want to say that shock is a real thing. And it makes us do strange things and respond in different ways. I say that because I interpreted their action as one of leaving me. Right or wrong, I raised my rifle and fired one round over their heads. I was unable to get back up on my own, the three of them had left me on the ground. After the shot was fired, the three of them came over and sat down next to me. Quietly we waited, not knowing if anyone had heard the radio transmission requesting help. Without anyone hearing the mayday call we would be on our own. After about 20 minutes we could hear the whop, whop sound that only a Huey makes. A few seconds later a sister ship that heard either my or the AC's distress call came into the field. They put the aircraft into a circling approach and landed about 150-200 feet away. There was a short Major that came running across the open area, grabbed my M-16, and gave it to the AC. Then he picked me up and put me on his back and went running back to the aircraft. The sister ship had my best man WO-1 Walt Melson from my wedding and WO-1 David Peters flying. I couldn't have been happier to be off the ground and in the air again. The relief that I felt exceeded the pain of the bullet wound.

> From the U.S. Army Goldbook
> Helicopter 65-12895
> Information on U.S. Army helicopter UH-1D tail number 65-12895
> Date: 670401
>
> This was a Combat incident. This helicopter was LOSS TO THEATER This was a Logistics Support mission for Re-supply, to Forward Area. (No it wasn't, it was to pay the soldiers in the field).
> Unknown this helicopter was on Take-Off at 0075 feet and 025 knots. South Vietnam
> According to the Major that went back to the helicopter and got me the left pedal and picked up several spent projectiles, he said, "The helicopter took over 80 hits"

The Army Goldbook said that the aircraft had 12 hits from: Small Arms/Automatic Weapons; Gun launched non-explosive ballistic projectiles less than 20 mm in size. (7.62MM) The helicopter was hit in the Cockpit, fuselage, Cabin area, Engine compartment.

Systems damaged were: HYDRAULIC SYS, OIL SYS, ELECTRICAL SYS, TRANSMISSION, COMM SYS, ENGINE, PERSONNEL, FUEL SYS Protection of PERSONNEL by Armor was Not Effective Casualties: 02 WIA

The helicopter Crashed. Aircraft is later recovered by any means other than its own power.

Both mission and flight capability were terminated.

Publisher's Note

I am Bob Babcock, the 1LT Babcock who Don delivered to the field on April Fool's Day, 1967. We did not know each other at the time of the incident, but now we are great friends. Following is the story of how we finally met.

In the spring issue of the 1993 Ivy Leaves, newsletter of the National 4th Infantry Division Association, I saw a note in the locator section: "Does anyone know who the intelligence officer with a brown briefcase was that I dropped off in a Montagnard village on April Fool's Day 1967? As I was leaving the LZ, I was shot down and wounded." He included his name and phone number. In those days before email and cell phones, I went immediately to my home phone and called him.

"Don, I'm the guy you are looking for. I wasn't an intelligence officer, I was the pay officer. Even though troops had no place to spend their money, they still expected to be paid on payday." We were thrilled to have found each other. We talked a little longer and decided to get together.

Each summer back then my family went to Panama City Beach, Florida, for a week at the beach. Dothan, Alabama, where Don lived, was directly on our route. In July 1993, Don and I finally met in person the guy we shared our April Fool's Day story with. As we visited in his home, he showed us the rudder pedal from the helicopter, with a jagged hole where the bullet went through before hitting his foot. Don's wife,

Joyce, and my wife and kids listened in awe as we talked about that most memorable April Fool's Day. They all knew about the story and were as pleased as we were that we had finally linked up.

We stayed in touch, met again at a 4ID reunion, and I found out about Don's role during the 22 March 1967 battle of Alpha Company, 1st of the 8th Infantry, where First Sergeant David McNerney earned the Medal of Honor. Don and I had both become friends with David McNerney, and had been accepted as honorary members who were invited to all the events their company had over later years. It made me appreciate Don even more as I learned from the guys on the ground what he had done to help them during that fight and earned the Distinguished Flying Cross.

A few years ago, I started encouraging Don to write this book. He initially (like most veterans) said he had nothing to write that anyone would care about. As an author who has written his own book about my Vietnam experiences, and now a publisher, I kept pushing him and finally he agreed to write this book. I am proud to be the one who is making it available to the reading public.

Following is the story I wrote, before I met Don, that is in my *What Now, Lieutenant?* book...

April Fool's Day, 1967

Every soldier wants to be paid on payday, whether he has a place to spend it or not. April 1, 1967 was no exception—it was the first day of the month, so payday it was.

The company had been operating for the past several days in an area northwest of Pleiku, a relative picnic compared to the jungles we had been working in since the 3rd of November. The terrain was fairly open, it was sprinkled with a series of Montagnard villages, and the only enemy seen were a few bands of Viet Cong.

It was a welcome relief to be out of the jungle, off the Ho Chi Minh trail, and fighting small VC groups instead of NVA regulars. Being

INTO THE STORM

payday, the company had not even sent patrols out. We were taking the opportunity to lay back, take it easy, and enjoy a well-deserved break.

The sun was shining brightly and the countryside was beautiful as the helicopter I was on approached the landing zone beside the company defensive perimeter. The Montagnard villages dotting the area had always intrigued me. I was enjoying the view and the cool breeze blowing through the open doors as the pilot radioed Captain Ator.

"Oscar 6, this is Black Jack 895, mark your position, over." Harry Troutman, Captain Ator's radio operator responded back in his familiar southern drawl, "Black Jack, this is Oscar 6 Echo, our position is marked, over." "This is Black Jack 895, I see purple smoke, over." "This is 6 Echo, roger purple, bring it on in, out."

As the helicopter started its flare to hover onto the LZ, I grabbed my M16 and my brown briefcase full of MPC (military payment certificates) and prepared for a quick exit from the chopper—pilots don't like to stay on the ground in a potentially unsecured area. As I jumped to the ground, the pilot accelerated and headed back up into the sky as I casually started to walk toward the company.

The sound of gunfire punctuated the calm as bullets raked across the helicopter. The chopper lurched forward as the pilot lost control. The copilot reacted quickly, regained control, and nursed the crippled machine to a hard landing not too far from the LZ. At the sound of gunfire, I quickly sprinted to the safety of the company perimeter (it would have been interesting to have a stopwatch on me to see how fast I covered that ground).

Bravo Company reacted quickly to the gunfire. Before the rotor on the chopper had stopped spinning (actually it broke off with the hard landing), a squad of men had formed a skirmish line and was moving across the landing zone to find the VC who had fired the shots. Another squad quickly grabbed their steel helmets and rifles and headed in the direction where the helicopter had gone down.

As I ran up to Buck Ator's command post, he was already on the radio alerting battalion headquarters. As I dropped down beside him, out of breath from my sprint, he asked me, "Did you see where they were firing from? Did anyone get hurt on the chopper?"

"Negative, I did not see anyone. Beats me if they hit anyone, I got the hell out of there as fast as I could."

Buck started issuing orders. "Harry," turning to Harry Troutman, "go check on the crew and see if they are okay and report back to me. Babcock, get some platoons sweeping the area and see what they can find. I am going to see if I can get one of those jets circling up there to drop his load for us."

As Buck started calling the Forward Air Controller to bring in an air strike, I called the three platoon leaders to alert them to get ready to move out. In less than five minutes, my old third platoon and the second platoon were started on a pincer movement around either side of the landing zone. They headed down into a wood line that led to a creek where we suspected the shots had come from. It was a lot more fun being on the command side telling the platoons where to go. In the past, I had always been out there slogging through the woods.

It became a game of cat and mouse as the platoons radioed back they had found nothing. Since I had a good view of the terrain, I kept adjusting their movement to try to get them to spots which looked like good hiding places for the sniper.

In the meantime, Buck had been successful in getting a flight of F-4 Phantom jets to blast the downstream side of the valley to try to cut off that escape route. Four jets screamed in, one after the other, and dropped their loads of bombs and napalm. As usual, the display of firepower was awesome.

After about an hour of futile searching, we called the platoons back in to the perimeter. A medevac chopper had come in to take out the pilot who had been shot in the foot and the door gunner who had been clipped in the nose by a bullet. Another helicopter was on the way to lift the disabled chopper out.

INTO THE STORM

I paid the troops, took notes on the pay and personnel problems I needed to work on, and spent the afternoon talking to the soldiers while waiting for the evening chow chopper to come in. Patrols were sent down into the valley before this chopper started the same approach we had taken that morning.

No shots were fired as it landed. As I climbed aboard for the return flight to base camp, I remained tense until we were well clear of the sniper area and had climbed out of range of small arms fire.

The next day, as the battalion commander and sergeant major were landing at Charlie Company's defensive perimeter, a single shot felled their helicopter as it was landing. No one was seriously hurt but the repercussions were widely felt.

The order was given, "A company will secure an area 500 yards all around a landing zone before any helicopter will land." This was not just an empty order; the company commander had to personally tell the pilot he had secured the 500-yard area before calling the chopper in.

Trying to secure a 500-yard area on a parade ground or parking lot might have been easy, to do so in the terrain we were in was impossible. Regardless of how much the CO's argued, the order was not rescinded. They were told they would be relieved of their command if another helicopter was shot down. Each time a helicopter approached his area, the company commander would cringe and worry about his career until the chopper was safely gone. (What a helluva way to fight a war).

Three days later, a helicopter was again fired on as it approached Charlie Company's landing zone. This time, a patrol was lying in ambush along the glide path. They opened fire on the Montagnard farmer as he fired, killing him instantly. He had wreaked much havoc on our operations before we finally got him.

***

I did not know until I edited this book about the tough time Don had in the medical system after he was wounded. I finished my tour in early July 1967, got out of the Army, and assumed that Don had recovered and might still be flying in Vietnam as I returned home.

I often think how just over a week before April Fool's Day Don had earned the Distinguished Flying Cross during a Medal of Honor level battle, and during the night before he picked me up he had been shot up while flying base camp protection. On those two, and other, occasions, he came out without a scratch. Then, on a seemingly routine flight to drop me off to pay the troops, he got a bullet through his foot.

Read on about Don's trials and tribulations recovering from his wound on that fateful April Fool's Day in 1967.

April Fool's Day has a meaning today that I'm sure is not shared by many. How did the rest of the day go one might ask?

Distinguished Flying Cross Ceremony

8. Medical Evacuation to Rear Echelon of Care

I WANT TO MAKE CLEAR AT THIS POINT THAT I CONSIDER THE BULLET WOUND I received a potentially non-serious wound that should not have prevented my return to Vietnam after recovery. It was pointed out to me in the beginning in the hospital that it was believed that I could be back as soon as six weeks. So, what happened we might ask, read on.

EVACUATION ITINERARY:

An interesting note about military recordkeeping, my DA-759 (flight-records) shows me flying several days beyond my being shot down and wounded. Now that would have been a neat trick, however, it didn't happen. I have quite a few examples of how documents were typed with "disappearing" ink. Some of you will know what I'm talking about. How many of you didn't receive the original paperwork but a "pink" copy made with a ribbon? I still have some of those and since they represent the only true and authentic records, they are worthless. I have enough of them that I firmly believe that our personnel division was aware that ultimately if we ever needed them to use, they would be worthless. Obviously, I received too many of them for me to feel the way I do. Anyway, moving on...

Triage is the process of determining the priority of patients' treatments based on the severity of their condition. This rations patient treatment efficiently when resources are insufficient for all to be treated immediately. The term comes from the French verb trier, meaning to separate, sift or select. Triage may result in determining the order and priority of emergency treatment, the order and priority of emergency transport, or the transport destination for the patient.

1 APR 67, 4TH ID PLEIKU, RVN 6TH EVACUATION HOSPITAL (DRAGON MOUNTAIN)

I was flown into the base camp and taken to the battalion headquarters, where I sat, on the CQ's desk, for an hour waiting for the ride to the Evac hospital. Bleeding all the time, I'm waiting for someone to come and get me. As I sit there, I pull my bootlaces as tight as I can physically. I am trying to use the laces as a tourniquet to slow the bleeding. I'm watching this little stream of blood just bubbling up and spilling over the side of my boot and onto the floor. Now I realize, I'm not going to die from this. But one still wants to get somewhere to get the ball rolling. I don't think I'm being unreasonable. Since I'm alone, I start looking around the room to see if I could find something I could use as a crutch. If I can locate something, I can get outside and there should be a jeep. I could stop and find someone that knows where the hospital is located. About that time one of the clerks walks in and asks what I'm doing sitting on the desk. "Just bleeding out," I said. "Can you give me a ride?"

Finally, I get to the hospital, where they're not that excited to see me. They wheelchair me in. It is apparent to me they think I shot myself. Not in a big hurry to help me. Do you remember the definition of Triage? Now it has nearly been two hours since I was shot when someone walks in and sees me. They recognize me and ask why I'm sitting there? Not waiting for an answer, he says he's a helicopter pilot that was shot-down a couple of hours ago and was wounded. Why is he not in surgery? I thought what a good question. By now I'm starting to get a little groggy. They take me to an open space, put me on a table, and begin cutting my clothes off and cutting my boot off.

Jungle boots have a steel shank in them to protect soldiers from stepping on something that would penetrate the boot and foot and disable them. In the case of an infantry soldier, it would help in the event they stepped on punji stakes.

In my case, if you look at the pedal, you can see the bullet went through sideways, entered the bottom of the sole, penetrated the plate mostly sideways, and exited my boot sideways. That's a hole the size of a half-dollar in diameter. The plate shattered somewhat and then opened up in a flower pattern inside my foot, destroying the bones, veins, and nerves. Bullets are very hot, friction from the air and striking surfaces make them so. This in some cases requires that dead meat needs to be trimmed back until healthy meat with blood supply is available. Or so the Doctor explained the debridement process to me.

Now I digress to the table, they seem to be working more urgently now. Got me naked now. They roll me on my side to give me a spinal they are having a problem with it. (Years later, I had an x-ray that shows no disc-cartilage between L-4 and L-5 with L-4 pressed against my spinal cord.) Finally, they stick it somewhere and roll me back. Now, this is funny! If you've ever had a spinal, you may know what happens next. My left leg is bent and my foot flat on the table. The surgeon says put his leg down flat on the table before that spinal takes effect. Well, they didn't. When it does take effect, your brain remembers the position the leg was in. Now when you wake up you keep desperately trying to put your leg down but it's already down. So how does one finally resolve the situation—time. My back was killing me. But since it wasn't bleeding, I guess they didn't care about that. I mean I just fell out of the sky and crashed all over God's half an acre and they weren't interested because there wasn't blood pumping out of my back. I was pretty doped up for about three days. Major that picked me up and carried me to the helicopter came in to see me. They brought in a couple of the bullets and the pedal from 895. He told me that bullets were laying everywhere, all that he had to do was bend over and pick them up. The chaplain gave me a tape recorder to make a tape to send home. The batteries were weak, so Joyce couldn't understand the tape. Stayed here a couple of days and then was transferred to the coast. Saw one fella that was hit with CBU (friendly fire). I think he said he was in Cu Chi. He looked like Swiss cheese. I admired him. He wasn't letting it keep him down. He was walking around. Very tenderly I might add. Stayed pretty doped up here—better enjoy it, because it was going to stop. Things are going quite well considering everything.

5 APR 67, 18TH SURGICAL HOSPITAL

Evacuated from Surgical Hospital, now this is interesting, so far, they haven't done anything here other than jam some gauze in the hole, slide gauze between my toes and wrap two miles of gauze around my foot. Joyce, I'm ok, check out the picture of the foot. A piece of cake! Cool, I can see through my foot. Hmmm, we got mortared tonight it really shook all of us up. Some of us were on the floor before we knew it. Sad one of the guys low-crawled to the end of the room. Both legs had been amputated. We needed help to get back into bed. Having trouble going to sleep after last night. If they attack, what do we do? I don't feel very safe here. It has been almost a week, wish someone would talk to us and let us know what is going on.

Ok, they came in and said we're going to Japan. Outstanding, we're going to Japan. We're looking forward to our evacuation to Japan. They've got us all on stretchers and we're on the tarmac for loading on the plane. Just before they loaded me on the plane, a Swiss doctor came over and saw the gauze in my foot. The dried blood and scab had formed in and through the gauze. He reached down and ripped it out. I sat up and passed out. When I woke up I had new gauze in the same spot. What was the point behind this? Had they screwed up? What was the point? I thought these folks were on our side. Don't believe it for a minute.

9 APR 67, CLARK AIR FORCE BASE PHILIPPINES — OVERNIGHT STAY IN US AFB HOSPITAL

I want to say that at this point what's the state of my mind. To say the least as I look back upon that point in my life, I am self-absorbed, prideful, feeling sorry for myself, angry, and self-centered. At the same time, there is a paradox between how I feel and what I see around me. My brain and emotions are totally at odds with each other.

Consider with me what is my wife of six weeks thinking as she reads my letters?

10 APR 67 TACHIKAWA, JAPAN — DISPERSING AREA

I was out cold when we were rolled off the bird into the hospital in

Tachikawa. I became aware of people talking and standing very close to me. I was startled and jerked awake. As I looked around, I noticed something shiny on my pillow near my left cheek. Someone had pinned a Purple Heart on my pillow. I was trying to focus on who was standing near the bed on my right side. There were two men and a woman. A photographer was taking pictures of the four of us. Someone then leaned over and said that Peter, Paul, and Mary were here on a singing tour and were visiting troops in the hospital. Ok, this is cool, I wondered if they were going to break out in a rendition of "Puff the Magic Dragon". I was in a publicity shot with them. Well hot dog, I'm famous now.

Guys in uniform began to move us to the loading area. Obviously, we're not staying here. They loaded us on small buses or ambulances with stretchers on them. Ours had three stretchers on each side. We took off and slowly traveled off base and out of the city. As we drove along and left the hustle and bustle of the city, the countryside gave way to some beautiful scenery. We came to a stretch of road that was straight, with farming on both sides. There in front of us in stark contrast with one another was on one side of the road a farmer using an ox and on the other side a farmer using a tractor. Then there was a moment when could see Fujiyama white capped with snow. This is going to be great. Please note I'm smiling — picture went to Joyce. We're going to Johnson Air Base. It seems during World War II it was a part of the Imperial Japanese Air Force. The US has occupied it for a while off and on. It appears that the Japanese are using it now. Supposed to be an airshow here sometime this year. Well, I shouldn't be here very long just until I get healed up. I've been told that typically you get sent back, possibly even to your unit you left. Coming in the gate.

JOHNSON AIR BASE, JAPAN — 7TH FIELD HOSPITAL

The wards date back to at least the Korean war. I don't think the Army has the manpower to operate their hospitals. They seem to be short-handed. We can go days without much interaction. One of our nurses is Japanese and doesn't speak English. Difficult to get anyone to get you any pain meds. Since I've got a lot of nerve damage, I'm lucky that I'm not in much pain. But some of these guys are hurt really bad. The ward is like a finger off a backbone. We're sort of out of sight and evidently out of mind. Well, this won't mean much to anyone but one 65mg Darvon was all you could get for pain. About the equivalent of two aspirin. That just isn't enough for some of these guys. Well, today someone finally came on the Ward. Went around checking some of the wounds and went back out. Someone came in a little later and did something I can't believe. He unrolled what looked like a napkin and removed something shiny. We weren't paying that close of attention. Until we heard this blood curdling scream. We all sit up and look in that direction and the guy is cutting him with a scalpel. It seems cold turkey surgery was normal and nearly an everyday thing. This place began to really mess with your mind. I am being told that I need to have surgery once more. That I'm having some infection issues and they need to fix it. Sounds like a plan to me. Faster I get healed; the quicker I can get out of here. The time comes and off to surgery we go. Now what happens next is a mystery to me.

I wake up and they tell everything went great, in fact so well they were able to close the wound and put me in a thigh cast. That sounded good, however, where a big hole was there was now a cast. What! It no longer needed to be cleaned and drained? So, a couple of weeks go by, and I'm running a fever and there is this smell coming from my cast. Upon closer inspection, I notice my skin is sort of a bluish green at the top edge of the cast with a black, blue, red, yellow streaks coming out of the top. My attitude quickly goes into the toilet. I finally grab someone and say I think something is terribly wrong here. She looks at it and says something, "Oh my."

She disappears and returns minutes later with a gurney and some help to get me on the gurney and off we fly down the hall. They run me into the cast removal room. Take their little vibration cutting tool to slice open the cast. They get the cast off and remove the bandage. The Doctor is sitting there and looks at it. Ok, you probably don't want to picture this, but my foot is about twice the size it should be. The bottom of the foot appeared to have closed. The top, where the bullet had exited, was the size of an orange. Now, consider that a moment ago I had a cast on. So, two things 1) the cast was stuck to my skin and puss had hardened in some spots, 2) the puss in my foot had been somewhere and now the pressure was moving it to the weakest part of my foot. It was expanding under the sutures, blowing up like a balloon. The Doctor paused a moment or two. He had a pair of suture scissors in his hand. I figured I knew what he was about to do. He reached down and put the little cutout piece of the scissors under the suture and cut.

Now the only thing I can compare this to is Old faithful at Yellowstone Park. When he cut that one it squirted out for what seems like a minute. He's trying to get away from it as it squirts on him. The smell is worse than week-old roadkill. Several of the suture holes, now as the pressure begins to subside, begin to slowly open-up. Some of the skin is rotted and flesh under the skin rotted. The liquid leaking out was a mixture of yellowish green and black. Ok, I'm sort of a novice to the medicine thing, but this doesn't look good to me. Doc what do you think. He looks at me and says to the nurse clean him up, give him some penicillin, and get some maggots.

On the way back to the ward, she says we have to move you off that ward. Why I ask, because you have an infection. Ok, and what was that about maggots? Well, we have special medical maggots that help to clean up infected wounds. Now at this point, as they are putting the maggots down inside my foot and the top out of curiosity I sit and watch them. They were very busy little critters. All this time I was thinking who do you suppose raises maggots for medicine, and how much do they cost? Could I start a business raising maggots for hospitals? They sure were hungry. Oh, I get some fresh ones today, lucky me. I'm laying here today thinking, wasn't I better off before I got here. Just what caused this? I had been sterilizing my hands and anything else that touched my foot and cleaning the wound three times a day. Then I began to think about the surgery that I had. They sutured the bottom shut then they sutured the top shut. What exactly was between the bottom and top sutures? The answer! An empty hole with no place for anything to go. Seems we have some very young graduates with little to no medical or surgical experience. What more can I say? Let's hope they're doing a better job on the more seriously wounded. Is this a fluke? I'm done with the maggots. Good, I've gotten tired of them. They kept falling off in bed with me.

Now I am on the infected ward. Hey, I can get my own room here. Well, I'm cleaning my own wound again. Phisohex soap, saline solution and then a treated wick to help drain the infection three times a day every day, it may be painful, but it's tolerable. (Right: The smile is for you sweetie).

I was told today that if they can get me healed up within the next month or so they can get me back in the fight. Sounds better than being here.

I am in the hospital with John Mayher, he was Outlaw 17 when COL Dempsey made his ill-fated charge into an LZ and died in the attempt. He was going in for the crew of Outlaw 17. John and I were in adjoining rooms. I could hear them doing cold turkey surgery on him every few days. His wound was closing too fast and they had to keep it open. You had to be careful how fast you were healing from the outside in. If it was too fast they would just grab a scalpel and cut you back open. No, they didn't deaden it for you before they did the dastardly deed. We had a newbie on the ward and some mysterious person walked onto the ward. Someone claimed he was a doctor. But we knew that couldn't be true. What he did next just didn't seem right, without a glove, he jams two of his fingers into the hole in the newbie's leg and probes around looking for shrapnel. Two orderlies held him down while the "doctor probed" around. Heck if this is what it takes to be a doctor, then I should have gone to Medical school.

Got a lot of crummy memories from here. Now they've decided they can't get me healed and back into action. The nurse said that it has something to do with the gangrene. They indicated that I might lose the foot above the ankle when I get back to the States. What is this? The Civil War? Yes, and I'm at Andersonville. Years later I spoke with MG Spurgeon Neel (Retired), otherwise known as the Father of Army Aviation Medicine. He told me about an article that he published. It seems that the best medicine was in-country and in places like Germany. He also said that High Level Commanders in Vietnam didn't mind having soldiers evacuated all the way back to the States since they had a large pool to draw from. In other words, it turns out they weren't that concerned about the soldiers. I have the papers that MG Neel published. Supposedly, the infection is under control.

18 MAY 67, TACHIKAWA, JAPAN — DISPERSING AREA

Thank God, I'm leaving.

19 MAY 67, SCOTT AFB, MO

Bad weather forced the Air Force C141 to land in St. Louis. More than likely someone simply was based out of Scott Air Force Base and had a date with an Air Force nurse. A "DC-3," you have got to be kidding me. Why not just hang us under the airplane and fly over at 3,000 feet and use the hospital as a drop zone! The Air Force loads us on a DC-3 and proceeds to beat us to death for eight hours. A DC-3! Come on people, are you kidding me? Here we go, they're delivering us to different hospitals in the Southeastern United States. I guess they're trying to spread us out. Hot, sweaty, noisy, no food, no water, and a long flight from St. Louis to Ft. Gordon, GA. Sadistic pieces of crap! This just keeps getting better and better. What did I expect, first class limousine service no, but at least the deuce-and a-half would have been more comfortable?

20 MAY 67, FT. GORDON, GA — HOSPITAL CORRIDOR A, WARD A-20

This was all so negative for me. Ok, so I arrive here thinking that things will be so much better here. I'm in the States, should have the best healthcare, the best of everything, right? Wrong! I know that some of you are thinking, "He's just writing to be writing this." Not so, Local Butcher at this location is COL Watson (Orthopedic Doctor). The wards are so full you can't get between the beds without moving the beds over. It is so crowded we are almost in bed with each other. The guy to my left is semi-conscious most of the time. His right leg is gone above the knee. His left leg is in various stages of rot. He apologized for smelling so bad. They came by and told him they couldn't save the other leg and that they were going to amputate his left leg and it will be near the hip. They haven't even been in to see him in a week.

Just how crowded is the ward. You must start at the end of the row in the ward and move the bed against the next bed to physically get to the patient. I am continuing to clean my own wound. It stinks so bad from the rotting flesh. You don't eat, if you do you'll just throw it up. Well, it's been decided to amputate my foot. This knife happy creep just wants to cut it off.

Some of these guys are just giving up and others are near violent. My God, how am I going to deal with this crap? The days are stretching into what seems one long nightmare. Guys are crying out, groaning and moaning all night long. Sometimes, what I think is someone else is sometimes me I'm hearing.

Can you imagine what it's like to wake up to the sound of a bone cracker? It sounded like a gunshot. We all are upset with this rocket scientist. The doctor decided to remove one of the bones in this guy's foot on the ward. Let's see, no sanitary anything—on the infected ward, for crying out loud. The kid on the right of me has a leg that is rotting off. It's as though his skin is being eaten by something. He said the doctor was going to remove the leg as soon as he got a little better. He died two nights later. I'm going to die here. There are purportedly over 8,000 of us here in this hospital. About 4,000 were assigned to the psychiatric wards and the rest of us were assigned to the orthopedic wards. Ok, I'm over my hissy fit. I'm trying to have a solid perspective on the situation.

The dental folks on this post pulled a trooper's tooth and two weeks later he was dead from infection. I wish I still had the newspaper article on this one. This kind of quality health care at home is totally unforgivable.

My wife came to visit me and was verbally abused by a cake eating solder in the billeting office. Once he found out she was an Officer's wife the tune changed dramatically. Think I was upset—what do you think? Joyce came on the Ward and it was so bad she couldn't handle it. She drove 12 hours just to see me for a couple of minutes. I couldn't blame her. It was an unbelievable, inconceivable environment. Lives destroyed by the enemy then further insulted by those that should have cared for us. Again, I hear that the casualty rates are so high that the Army isn't prepared to handle them. From my perspective, I would say unprepared would be an understatement. Same is being said of the psychological cases. Either they transfer me out of here or I go AWOL. I'm going to call for a taxi and "walk out." I've had all I can stand. Joyce has begun working on a compassionate transfer to Ft. Rucker, Alabama. Seriously, I'm going to call a cab and leave.

INTO THE STORM

As it turns out the threat of the Congressional investigation was all that was needed. I'm being transferred to Ft. Rucker Alabama.

25 AUG 67 FT. RUCKER, AL — LYSTER ARMY HOSPITAL

LTC Robert L. Reid, MC Ortho Doc.

I'm in the old-World War II hospital. But what the heck, I'm home. They've just built a new hospital on the hill. They don't want me where the flight students can see me. Seems I'm the only wounded Nam vet here. They don't want my condition to be a negative influence on their training. That's ok, I don't want it to be either.

Joyce and I can see each other now. That aspect of pressure is off both Joyce and me. Doctor Reid formulated a treatment plan; however, he is sure we will have a successful outcome. He confirmed that the condition of the Achilles tendon and the bones in my foot were not real positive. I am so excited. Someone has a plan! Successful or not, we have a plan. He explained that he was going to stretch the tendon. He was going to place me in a full leg cast but cut part of the top of the cast back to where I could keep cleaning the wound. He was going to put me on convalescent leave for one week and then come back in for evaluation. I thought I had died and gone to heaven. When I returned from the first week, he said there was some improvement in the X-ray and that whatever I was doing to keep it up. He sent me back out for another week.

13 SEP 67 EVALUATION

Dothan Eagle called and wanted to do an interview with the both of us. So, we said sure come on. Mary Bubbett was the reporter. She took the picture above and wrote an article called "April Fool's Wound — No Joke".

Joyce and I went to Panama City to Petticoat Junction. She and I were going to ride the train when a car backfired. Took about 10 minutes to calm me down and out from under the train. I weigh about 104 pounds so I'm really skinny, about 22 pants. Just some bone with skin on it. But I'm the happiest boney skinny guy in the world right now.

20 SEP 67, EVALUATION

Dr. Reid says that the x-rays look very promising, however, not out of the woods yet. He is cautious because he has told me that amputation is not yet off the table. "Mister Rawlinson, you are about to start a "tough" regiment of physical therapy." What does "tough" mean? "Meet THE Major and head of Physical Therapy. She served during World War II, Korea, and now Vietnam. You will probably be her last patient, so I expect you will recover to a level of 100 percent." Now she looks like my Grandmother so that can't be too bad? My education continues to expand to all different levels in life. She introduces herself and says, "There is no rank in this room. There are no excuses. You WILL meet my goals on your own or I WILL assist you in meeting those goals. We start today." There is no getting around a woman that is committed to her job. Before this adventure, I was injured, angry, mad, furious, however, now I'm flat out scared.

DAYS OF PHYSICAL THERAPY — REPETITION

Each day I enter that room, feel as though I have entered Helga's House of Pain. "Move that foot," she said. "Do you need for me to assist you?" "No! Please, I can do it myself!" She grabs my foot and begins to make the ligaments and tendons stretch. Ouch, help, bam, bam as I pound on the table. Ice, heat, more stretching… days move into weeks. Where did she go to school? The Olympics Wrestling team would have nothing on her. Over the weeks, I realized that she had only my best interest at heart. If I was going to be her last patient before retirement, SHE wasn't going to fail. Physical rehab was truly torture. When it was over, I told her that she probably saved my life. However, if it weren't for her and the orthopedic doctor, I'm sure I would have lost my leg. He had a novel approach to working with me. Classic case of I don't know what to do for you so go home, take two aspirin, and call me in a week, only he had a plan. Now, this was humor I could deal with. He was watching to see if signs of life were present, both in the injury and the patient.

Looks like the Doc may set me free.

I received a Disposition Form (DF) on 26 Sep 67 directing me to attend a debriefing of Vietnam veterans at Building 5205, Room 6 on 18 October at 1300 IAW USAAVNS regulation 350-23. They have decided they will tell me how and what to think. Ok, I need this, really, I do.

9. Try to Find a Way Out of the Storm

OCTOBER 18, 1967

Psychological evaluations followed and lasted about two weeks. All it amounted to was brain washing to make you fit back into the world. I answered the questions that they were asking. They didn't like the answers and I didn't catch on immediately. Finally, I understood I was being asked the same questions different ways. Then once I began being less than truthful, things seem to start going forward. It was just a matter of telling them what they wanted to hear. At this point, it was a done deal. Bodies nearly healed, and the brain is ok! Body yes, brain not so much.

This is an emotional spirited discussion due to the inappropriate behavior of the medical personnel here in the States. Both military and civilian behavior was totally unacceptable. I was more angered by the attitude of civilian employees and "cake eater" soldiers than the behavior of the public. The truth is that many of us are not brothers or comrades in arms. Every individual must prove their position "Deed Not Words". I will forever be watchful of what a person does and not what they say.

This tragic time of my life has revealed so many flaws in the American people, political leaders, family, and friends. Joyce is my only reason for living. (The latter is how my bitterness forged my view of everyone outside of Joyce. I look back at this period of my life and I pity me for being the cause of many of my own problems).

I am beginning a spiral downward. I am trying to enter the "World" again and I am finding it very difficult. (Serious attitude and emotional problems trying to matriculate back into a society that didn't want you back). The Army and the civilian work force are filled with wannabes. Most of us have quit wearing our ribbons in silent protest of the attitudes and behavior of the military and civilian population. I feel I have deserted my comrades in Vietnam. I am alone plus one.

NOVEMBER 1967 FT. RUCKER, ALABAMA

Assigned with the Rotary Wing Plans Branch. I have volunteered for high risk aircraft and experimental testing with the Rotary Wing Plans Branch. Tested strobe lighting on aircraft in hazardous conditions. They have put some of the lights on the tips of the rotor blades. We went out to County Line Baptist Church on highway 84. On the property, there was a huge dust bowl. (Volunteers only) Tested experimental lighting under hazardous conditions (more dust). I felt I just wasn't ready to return to duty. I was physically sick and would vomit every day going to the flight line. Once I was flying, I would be alright. As it turned out, it was a good thing that I went back on duty. It made me push most of my problems to the back burner. I didn't resolve any of the issues, only buried them. Back to the science project at hand. The test was conducted in this manner. Two Hueys outfitted with experimental lighting sitting opposing one another with a huge dust bowl dividing them. The goal to be able to avoid one another by seeing lights in time. OK, we understand this is the ultimate game of chicken. We consider what we, the pilots, must do to survive. How do we make this a science experiment instead of a joke? As we look at the situation, the aircraft, and dustbowl we agree to the following: 1) We will coordinate our pick up to a hover, prepare to immediately go IFR in the dust. 2) The pilots will maintain the ground in view through the chin bubble. 3) Pilots will communicate when committing to entering the dust bowl, 4) Left seat to call our heading, 5) Left seat and pilot, as able, are to look for visible signs of lighting. We begin and everything was going to plan, until?

Assumptions were made or not considered. 1) What happens if we don't see one another? Is there a way to avoid impact with each other? If we see each other, how do we evacuate the dust bowl? We both are squeezing the trigger talking to each other.

"Do you see anything?"

"No, how about you?"

"No."

"What's your forward momentum?"

"Not fast, barely moving."

"Same here."

"Ok, I think we should be seeing lights by now."

"Me too."

"Do we call this quits?"

"Suppose so."

"Can you set down?"

"Yes, I think so—Yes, I'm on the ground."

"Ok, I'm setting down."

"Let's see just how close we are to each other."

"Works for me."

"Let's see who sees lights first."

"OK" I'm glad we stopped I don't see any lights on either one of us."

"Nope, sand must've eaten them, or we slung them off."

"How far apart do you think we are?"

"I would say about two feet."

"Well, I would call this one a bust."

"Me too—sit there I'm going to pull pitch and get out and away from the dust."

DECEMBER 1967

I must have killed some brain cells while there. Developed quite an alcohol problem here. Spent some time in the hospital due to alcohol poisoning. Took about a week to clean me out and get me back on flight status. You won't find that in my medical records either, I don't think.

Joyce and I are going to Michigan for Christmas, leave for two weeks.

JANUARY 1968

I've been assigned to the Method of Instruction Training (MOI) (More character-building exercises.)

INSTRUCTOR PILOT TRAINING

During this time, I was totally absorbed with Joyce, the Army, and flying. I had yet to deal with any of my problems from RVN. I would suffer from anger, depression, and nausea every morning on my way to the flight line. I was able to fly fine; it was just getting there that was the problem. I remember on my check ride as I picked the aircraft up to a hover, I could tell that the hydraulics had failed. I calmly announced to the SIP that I had to set the aircraft down. Then I told him that we had no hydraulics, he didn't believe me. Therefore, he tried to pick it up to a hover. I passed the check ride.

I had a problem with locked doors. Joyce locked me out of the house one day. I ripped the door off the house and threw it into the yard. I broke the windshield of the car with my fist one day. I wouldn't avoid fights and drinking continued to be a problem.

Now I was about to start training more of America's good men to fly in Vietnam. I wasn't sure that I wanted to do this, but I didn't care anymore. I did it anyway. I was a trained monkey on a string. I hate the thought that I'm so indoctrinated that I can't change a thing. During training, I'd let the students pretty much try and do maneuvers if they were game. We did engine out landings into confined areas. The army had called a stop to some of the critical maneuvers that students really needed to know how to do. So, some of us continued to teach 180-degree precision autorotation to a spot with zero ground run. You had better learn to do them. High speed 50 foot off the deck decelerations to autorotation. Confined area precision autorotation's and 40 knot tail rotor failure landings using throttle (no feet on the pedals) to steer while landing on the runway. These were some of the talents that an army aviator needed to be proficient in to fly in Vietnam, not just a normal landing and takeoff. It didn't do any good to demonstrate how to do a particularly dangerous maneuver without teaching it to them.

During my year and a half as a flight instructor, I trained more than 60 young men. Some came from Guatemala, some from the Navy, most from the Army. I was a Contact IP and proud of it. My job was to train students to fly the Huey. I was very strict and old at 22. I had one student at the beginning of my teaching that broke off a DF antenna. I used it as a pointer (at the instruments) and as an attention getter. I'd reach over and "tap" them on the shoulder with it. In every class, we spent one week in night flying and maneuvers. We taught night time approaches with and without lights. We would do night time autorotation's with and without lights.

My daughter Michelle was born in May of 1968, Joyce would bring Michelle out to the stage field and watch us fly at night. On one particular night, during a night autorotation, my student decided to pull the pitch at high altitude in a UH-1A model. I slammed the collective back down, but the damage had already been done. The rotor RPM was so low that I figured we were to crash for sure. As we fell, I pushed the nose over a little and the aircraft responded. As the nose lowered and the speed increased it would only be a few seconds before we would impact the ground. We barely made the runway when I flared as much as I could to build RPM in the rotor. Then I quickly lowered the nose and began to pull pitch for impact. We touched down at 80 knots and the tungsten shoes began to get hot. We had sparks coming off that went higher than the aircraft. I kept, lowering the collective to get us to stop without ripping the transmission out of the airframe. Smoke and what appeared to be flames, due to the magnitude of the sparks, were engulfing the aircraft. As we ran out of runway and inertia, the aircraft yawed 90 degrees to the right and the left skid fell off the runway. Smoke was billowing all over and we could barely see inside the aircraft. It takes the engine of an "A" model nine seconds to spin the turbine back up to operating RPM. Sitting in the smoke, in the background, I hear the turbine spinning back up. I learned that I could function in situational stress in aviation. But outside I had problems. Joyce was getting into the car with the baby and then drove away. She never came to watch us fly again.

    I soloed a student tonight. Put another student in the left seat. Always nerve wracking when you solo the students at night. You have confidence in their ability but there is always that doubt if there is a problem. So, I got out of the aircraft and told one to get in my seat. "Ok, now make three trips around the pattern and one autorotation." They're like my children, I watch as they take off and continue straight ahead and don't turn on the crosswind leg. I go running up into the tower. "Contact Army 12345 and find out what is going on." The student came on the radio and says he's experienced a hydraulic failure and that he is on approach to Highway 84. He is going to shoot a running landing onto the highway. Oh, brother, I can't watch. Someone let me know if you see a fireball. They land safety and "Safety One" flies them back to the stage field. One of the maintenance pilots is dispatched to fly it back. When he gets there, he states I can't pick this up to a hover. Get some ground handling wheels out here to tow it out from under the power lines. Now at this point as far as I'm concerned, he passed his solo flight. Good job!

Poster for "HONOR IN THE VALLEY OF TEARS"—The story of the men of A Company 1st Battalion, 8th Regiment, 4th Infantry Division on March 22, 1967

Mike Moran and I out at the flightline at "Flatiron"

Joyce and I

INTO THE STORM

My retirement flight and two of my favorite friends

DEPARTMENT OF THE ARMY

DONALD L. RAWLINSON

IS PRESENTED THE

SUPERIOR CIVILIAN SERVICE AWARD

for exceptional meritorious civilian service as the Information Management Officer, United States Army School of Aviation Medicine, Fort Rucker, Alabama, from May 1987 to May 2007. His outstanding professionalism, leadership, and technical skills resulted in major contributions to the success and cost-saving value of various critical programs. Applying extensive knowledge and experience, he aggressively pursued and resolved countless automation issues and led all efforts to reorganize the staff for greater efficiency. He personally directed several key actions, which made sweeping changes and significantly impacted the Command's mission. The distinctive accomplishments of Mr. Rawlinson reflect great credit upon himself, the United States Army Medical Command, and the United States Army.

6 June 2007

RUSSELL J. CZERW
Major General, DC
Commanding

A copy of my citation for the SCSA—It makes for a fun read.

DONALD RAWLINSON

Superior Civilian Service Award
Justification

Mr. Donald L. Rawlinson has distinguished himself with over 20 years of exceptional meritorious federal service. As the Information Management Officer (IMO), United States Army School of Aviation Medicine (USASAM), Fort Rucker, he exemplified the utmost professionalism and technical expertise, making him a leader in his field. He not only earned the respect, trust and admiration of his peers, but his opinions and counsel were actively sought throughout the U S Army. His dedicated efforts and astute knowledge of industry led to ongoing cost reductions in the procurement and implementation of USASAM's Information Management/Automation program. Mr. Rawlinson's contributions to the Aeromedical community are both imaginative and constructive, and have had a profound influence on USASAM and Fort Rucker. He served as a mentor to the staff by his examples and leadership style. The breadth and scope of Mr. Rawlinson's contributions have impacted every facet of this organization and the Army Medical Department (AMEDD).

Mr. Rawlinson's vast knowledge and expertise as USASAM's IMO was instrumental in the support and implementation of the G6 Chief Information Officer's (CIO) initiatives within the Army Aviation Medical Community. Despite a dynamically changing environment, he was able to resolve many complex issues. He consolidated the servers to eliminate costs associated with proliferation and maintenance requirements. He provided long-term solutions with existing hardware and software by matching the capabilities of the equipment to the user, maintaining a $0 growth. He designed, procured, and implemented USASAM's Computer room, which has been widely used as a template by other Fort Rucker agencies as a benchmark for success in Information Systems. He showed true innovation with his program to co-locate the US Army Aeromedical Activity (USAAMA) Division's resources for current and future mission requirements; an approach that is being considered for implementation by the US Army Aeromedical Center (USAAMC) as a potential cost saving initiative. His uncanny abilities to design, analyze, budget and implement USASAM's automation infrastructure has improved production and consolidated excessive automation hardware. He has saved the Army countless dollars and is truly a visionary in this technical field.

Because of his exceptional versatility, Mr. Rawlinson was able to perform a myriad of tasks. With the current Global War on Terroism (GWOT), he recognized the need for more Distance Learning (DL) opportunities and therefore wrote and distributed four cutting edge DL packages to the Army Aviation Resource Library for Initial Entry Rotary Wing students and AMEDD personnel. He also played a vital role in successfully designing three DL classrooms and acquiring the necessary Video Tele-Conferencing (VTC) systems for these classrooms. His efforts enabled USASAM to conduct real-time classroom training to forward deployed units around the world. These courses benefited numerous outside organizations and paved the way for the future 68W and Flight Surgeon Refresher Courses. Mr. Rawlinson was also instrumental in the conception, layout, and design of the Medical Suite Simulation Training building which resulted in a $1 million congressional grant for contruction. Additionally, he demonstrated outstanding leadership by

Copy of my citation for the SCSA

involving users throughout the selection, content development, and implementation of a new collaboration website, offering deployed personnel "reachback" capability and projection platform.

Mr. Rawlinson's work on the Operational Aeromedical Problems (OAP) Course continually provided the means for a highly successful Aviation Medicine conference. The OAP Course is an off-site, joint medical endeavor for flight surgeons which bridges the gap between current aeromedical issues throughout the entire Combat Health support (CHS) realm of responsibility. As the IT specialist and assistant planner for 12 annual OAPs, his extraordinary skills in the disciplines of audio-visual, computers, and networking, allowed USASAM and event coordinators to process information and requests more efficiently. His well executed planning and coordination saved thousands of dollars in outside technical support, which in turn helped to defray the costs of being able to bring in flight surgeons from the field to the conference. Because of his contributions, the knowledge gained by the flight surgeons during these events have truly been immeasurable within the CHS channels.

Between scheduled courses, Mr. Rawlinson volunteered a great deal of personal time in the Flight Simulation Center teaching, training and demonstratin to the students the basic of rotary wing flight. His exceptional technical skills of troubleshooting, applying innovative solutions to solve systems failures, and evaluating new softward packages provided lasting effects for a generation of officers and non-commissioned officers. His willingness to share new ideas with everyone was unsurpassable and he truly was a role model for resource sharing.

Mr. Rawlinson was no doubt dedicated to assisting the Department of the Army and the AMEDD Center and School (AMEDDC&S) provide quality Information Management Services to the aeromedical community. His professional efforts earned him the 2002 MEDCOM IM/IT Professional of the year award and he was named the AMEDDC&S Civilian Employee of the 3rd Quarter, Fiscal Year 2005. His contributions over 20 years of federal service have directly and positively impacted USASAM, USAAMC, USAAMA, the US Army Aviation Research Laboratory, the Medical Evacuation Proponency Directorate, and numerous other units throughout Fort Rucker and the US Army. He certainly deserves the recognition that the Superior Civilian Service Award conveys.

Copy of my citation for the SCSA

10. The realization that the Army has filed for "Divorce"

JUNE 1969 — THE DRAW DOWN

Unbeknownst to most people, the Army was going to start drawing down the number of Aviators on active duty. Many of us thought we would be able to get in at least 20 years and make retirement.

While at the stage field one day we were sitting in the field house discussing the flight and going over what the student needed to work on. The door opened and in walked someone from the personnel office on Ft. Rucker. They called us to attention and made the following announcement. All aviators in this building, if any of you would like to be discharged today and go home, please report to the personnel office on Post in the next 45 minutes. Several got up and then one of the enlisted drove them to the post. Several others simply got into the Huey's and flew back.

Due to the number of Helicopter Pilots that had been trained, over 40,000, the Army began offering early outs for its pilots. I tried everything I could to stay in the military. Contacting my Warrant Officer Advisor at the Pentagon. Volunteering for any branch that would accept me. Volunteering to return to Vietnam. One doc said I was good enough for stateside duty but not deployable.

I had a job all lined up with the resort at Lake Tahoe as a resort pilot. When I took my exit physical the FAA grounded me until 1974. Needless to say, I didn't get to keep that job.

11. Divorce is Final — "Hero to Zero" in a Signature Line

AUGUST 1969 — DISCHARGED 15 AUGUST 1969

We move to Decatur Street in Dothan, AL.

I applied for jobs with the FAA — air traffic control, Alabama Highway Patrol, Dothan Police Force, Houston County Sheriff's Department. I applied at Ft. Rucker for civilian jobs in flight training division, instrument training Facility. Turns out that a Federal Judge had hiring limited to minority hires only. So, I can't get a job. Got to do something or we'll starve to death. I can't get a job that I have experience for so what do I do? I applied for an insurance sales job.

1969-1970, SOUTHERN LIFE AND HEALTH

Worked there till about March of 70, turns out my route manager was stealing from me. I quit.

DONALD RAWLINSON

1970 — 1975, WALLACE COMMUNITY COLLEGE
(TALK ABOUT COMMITMENT)

Enrolled on the GI Bill to get an AA degree and then to pursue a B.S. in Business Administration

1970 — 1971

Worked at a Radio Shack and selling shoes at the Mall.

1971-1979, TERNES STEEL — SLIMFOLD MANUFACTURING

The company manufactured steel bi-fold doors and originated in Detroit, Michigan and was owned by the Ternes Brothers of Detroit. Evans Products purchased the Company and opened another factory in Dothan, Alabama of Slimfold Manufacturing. Primary products were steel bi-fold closet doors and a product for the automotive industry referred to as "Weariron." I was hired as the purchasing agent. Subsequently, up to 1979, I held positions in purchasing, inventory control, shipping, production management, and finally as the Materials Manager for the Company. During this time, the company was sold by Evans Products to a consortium called Dunbarton. The owner Nickolas Theofel had been the AT&T vice president over South and Central America. I enjoyed working there and with the people.

It during this time, that I continued my education at Troy University and my education in electronics.

1978

INTO THE STORM

I quit Dunbarton and took a position with Garwood Industries.

During that time, I purchased steel directly from the mills. In this job, I had to acquire chemistry and engineering skills. I worked closely with foundries and mills in the manufacturing and chemistry of the steels we used. I had the aptitude for it and I was in college and had already taken classes in chemistry, physics, math so it wasn't a reach to do the job. I rather enjoyed the daily challenge that was present. However, my personal problems kept getting in the way of work.

1979-1980, NORTON ELECTRONICS

I continued to find myself needing to be away from people. So, I quit a very good job. I now had my bachelor's degree in Business Administration and Computer Science. Guess what I did next, I became a radio and TV repairman. I locked myself away in the country repairing radios, televisions, and anything that came in. Electronic modules on cars, firing systems and ignition circuitry on peanut dryers, computers, workstations, monitors. It all just made sense how to repair so I enjoyed fixing things. I was asked one day if I thought that I could figure out how to expand the business and/or make it more profitable. I said yes. Rent the electronics daily and rent to own. So, I set up the rental agreements and people began renting everything from VCR's to big screen TVs. Almost everything had a payout of 30 days to six months. So, everything beyond the first week to six months was pure profit. Originally, I set it up to be floor planned with a local bank. Since we were horizontally and vertically integrated on a small scale, we could service, repair, or build everything we were renting and or leasing. He was getting rich. He commented one day where were all the people coming from. I said from as far as 75 miles away. No one else is doing this. Again, my issues get into the way of life.

DONALD RAWLINSON

1980-1981-1985, RILEY BUSINESS COLLEGE

I took a job really teaching computer classes. As I did more, I began to develop courses. The owner purchased two Model 2 Apple Computers that had two single-sided 160Kb floppy disk drives. He asked, "Ok, what can we do with these?" I sat down over the Christmas holidays and wrote several programs. Then I found a program that had been written, called the Executive Secretary. We can build an entire curriculum around this single program. We did just that. He purchased 10 or so computers and we began selling a package for six months for $1,800. As the computer industry expanded in 1982, I was working with an agriculture economist at Auburn University. The goal was to write a program using linear programming. I told him we could do it easier with a spreadsheet program initially. I used a product called AceCalc first then ported over to VisiCalc. The purpose was to take a product and allow the Agri-businessman (farmer) to input farming data and have the program determine the net-profit to land and business. Using a form of farming called no-till, I began to learn all there was to know about specific areas of farming. Such as how many kernels of corn go into the hole, temperatures and moisture for proper germination to encourage healthy growth, what percentage germinate, growth survival and projected yields. Well, the spreadsheet worked so well that we were able to capitalize the farm with 100% bank financing with 100% payback with amended profit projections supplied quarterly with the quarterly payment. It worked so well. I was put in charge of finances of one state senatorial race and then one federal level Senate race. Somewhere in all this mess, I had fooled around and picked up my MSc degree in computer science. Again I needed to get away from people so I left.

1986, WALLACE COMMUNITY COLLEGE — TOTAL 13 YEARS WITH WALLACE COLLEGE

INTO THE STORM

I was hired to teach various types of non-academic computer and office automation and corporate level software integration and operational software. Primarily we taught classes on new software being integrated into the workplace. This software required more than just operational understanding, it required understanding how to implement some of it into specific types of office environments. Data is no good if one can't perform accurate data mining and reporting. It was so very cool; I loved doing it. At this point, computers were making a difference in how the workplace was evolving. This was great I was an expert on desktop computers and had experience on IBM 360 and 370 mainframes, IBM 129 card punch, IBM 3741 and 3742 dual entry 8-inch floppy workstations. I had experience with different operating systems, software, programming languages. I was on a computer engineering team and possessed a Computer Engineer designation from IEEE engineering. Plus, I had maintained my standing with RCA and Sylvania as a recognized electronics experimenter. As I taught classes I began to get to know people. It was in one of those classes that I met some wonderful people that worked at the Morale, Welfare and Recreation Division at Fort Rucker Alabama. They asked me if I would consider working for the government. Now let me see—oh yes, I've wanted a job working with the government since I got out of the Army.

1986-1989, DIRECTORATE OF COMMUNITY AND FAMILY ACTIVITIES (DCFA)

You have hired the right man. I wrote over 120 programs and put in a network. Automated the club system, increased profits and lowered costs. I automated the Recreation facility, fishing and hunting permits. Created one of the first point-of-sales systems. Developed a financial spreadsheet that operated Ft. Rucker's Financial Resources Division. Automated Family Management Division then I was beginning to run out of things to do. I was on the wrong side of the house. I was in the non-appropriated funds section where there wasn't as much job security. So, I began to find a way to get onto the appropriated funds side of the house. I wanted to stay where I was but, unfortunately, a friend of mine already had that job.

I had put so much intellectual capital into the organization that I wasn't sure it would stand if I left. But I felt compelled to do something.

1989

Universal Energy Systems (Based in Dayton Ohio) Wright State University as a think tank

Senior Technical Manager

My job was to monitor the Commerce Business Daily and bid Federal contracts, to develop software solutions, manage contracts, to meet with prospective clients and discuss their problems and offer solutions through the use of technology.

Paravant computer systems — Manufacturer of Ruggedized MIL-SPEC 810-D equipment that was to be paired with 1553 databus analyzer for military equipment and civil aircraft.

MARTA — Atlanta to develop a software system capable of monitoring track maintenance and safety.

BARTA — San Francisco develops a software system capable of monitoring track maintenance and safety.

General Motors — Analog to Digital vibration testing analysis using hall-effect transistors and reed switching.

Sikorsky development of a new generation of helicopter

Cobro — automation development of Realtime flight time recording for maintenance. (Squat-switch technology to record actual flight time.)

1989, DCFA REHIRED ME

INTO THE STORM

No one could figure out how to keep everything running. So, at this one juncture in my life, knowledge was everything. After what turned out to be a six-month hiatus, I returned to my friends within Ft. Rucker. However, it wasn't the same. Things had changed and the work environment was no longer the same. So I spent several months training someone I felt could keep it all running.

1990-1991, US ARMY AEROMEDICAL ACTIVITY

Dec VAX and Dec Control Language (DCL) Datatrieve

Conducting Research and Data Mining on a system maintaining Aviator records and working projects on an as needed basis for the Director. Setup and maintaining offsite or remote data input locations.

1991-1995, US ARMY AEROMEDICAL CENTER

First Novell network initially 30 workstations growing to the entire hospital. Maintained 160,000-line programs developed to do patient scheduling. Developed additional 200,000 lines of code to expand the software into all functional areas of the hospital. Deployed a distributed computing platform consisting of 13 AT&T System V Unix systems running a MUMPS software system called the "Composite Health Care System" (CHCS) a total hospital integration system. Integrating pharmacy and automatic labeling and dispensing capability. Certified in AT&T System V Unix, Certified Network Administrator, Certified Network Engineer. It was Desert Storm that sucked me into the war as a civilian. Developed the graves registration software and repatriation software for returning soldiers at Camp Shelby, Mississippi. The Aeromedical Center at Ft. Rucker, Alabama was responsible for Camp

Shelby in Mississippi. Created a remote solution for patient-data exchange with Ft. Rucker.

Developed remote cost-effective data capturing for medical catchment area.

Installation of satellite star topology networking using Single and multi-mode fiber optics. Designed, developed and deployed campus network connectivity and management of medical resources to remote training facilities.

Deployed, administered and maintained a third-party pay/collection system

1996-2007, US ARMY SCHOOL AVIATION MEDICINE

Began all over again. All they had was a "Wildcat" bulletin board system and modem communication.

Deployed the first ISDN lines and fractionalized T-1 communication in the state of Alabama in 1996. Tele-videoconferencing and training worldwide by using state of the art technology.

Implemented school networking using Windows then progressing to Windows Active Directory. Distributed computer-based training software for classroom and web deployment. Developed computer-based-training software for field deployment.

Deployed Active Directory across training server-farm and maintained.

Continued to achieve cost reduction and focus on increasing the quality of training at the same time maximizing technology to reduce Total Cost of Operations.

12. So, what if Anything makes a Veteran Resilient

AS FOR ME AS I LOOK BACK TO MY EARLY YOUTH, I CAN ONLY SAY "FAITH" THAT God will make provision for me daily.

From the time I can remember my mother would play nursery rhyme 45 rpm records for me to listen. That lead to an understanding of who God is which allowed me to understand my place in life. At the age of 8, I came to know what God expected of me, however, I did not have an established relationship with Him.

It was through this process, however, that I learned a moral right from wrong. Even though I enthusiastically joined the Army with the understanding that I might be called upon to "kill", I had not embraced the true meaning of that understanding. Therefore, the closer I got to being shipped out to Vietnam, I began to have doubts as to whether I could kill someone. These doubts led me to consult with the local Chaplain. After several meetings and discussions, he was able to convey that killing had its place in society whereas, murder did not. There is a difference. Most of us do not see any difference, however, biblically there is.

From the time of arrival at Ft. Knox through the time at Ft. Lewis, I was as "lost" as anyone could be. In my mind and action, I had slipped into a period of moral depravity that I dismissed because I was a soldier and as such expected to do such things. To be quite honest B&B (booze and buddy) therapy every night was the standard way of life. Then throw in sexual relationships of the one-night-stand variety and one loses sight of reality in relationships. How does one find a real relationship when all they're looking for in the night is to forget the day?

Now fast forward to Vietnam. I speak now only to my experiences of that 90-day period. What governs how we perform our job? Since we are creatures made of three parts, physical, emotional and spiritual then consider the following:

1. Physical
    a. How much quality sleep does one get (minutes to hours)
    b. What kind of food are you eating (cold C-rations)?
    c. Work Environment (pilot, combatant, tanker, truck driver, etc)
    d. Climatic conditions
        i. Hot
        ii. Cold
        iii. Rainy
        iv. Foggy/Damp
        v. Altitude
    e. Chills
    f. Thirst
    g. Fatigue
    h. Nausea
    i. Twitches
    j. Vomiting
    k. Dizziness
    l. Headaches
    m. Rapid Heart Rate

- n. Muscle Tremors
- o. Grinding of Teeth
- p. Panic Attacks
- q. Chest Pain
- r. Tunnel-Vision

1. Emotional/Behavioral
    - s. Fear
    - t. Guilt
    - u. Grief
    - v. Denial
    - w. Anxiety
    - x. Agitation
    - y. Irritability
    - z. Depression
    - aa. Intense Anger
    - ab. Apprehension
    - ac. Emotional outbursts
    - ad. The feeling of being overwhelmed
    - ae. Loss of emotional control
    - af. Loss of control of bodily functions
    - ag. Questioning Authority
    - ah. Withdrawal
    - ai. Antisocial acts
    - aj. Inability to Reset
    - ak. Change in Speech patterns
    - al. Hyper-alert
    - am. Increased alcohol consumption
    - an. Survival Guilt

1. Spiritual
    - ao. Questioning God

ap. Questioning your own Faith
aq. + or—Change in faith practices
ar. Change in church attendance (yes, they have church in the field)

These are observed tendencies and traits during 1967. Many of these issues continue even 50 plus years after 1967. The question in my mind and the struggle has been to perform a mental "reset". A means of returning to my youth and/or "dumping" knowledge or exposures to things that I wish to erase from my memory. However, I know that is not possible. I have seen what I've seen and as redundant as that is, so it is so. I have read and studied why our brain functions the way it does and why my "reset" does not exist. No matter how I process the information I still reach conclusions that prevent me from "making sense" of the situation that is acceptable. As research suggests our amygdala, is shown to play a key role in the processing of our emotional responses. If then it cannot make sense of situations and diffuse responses, then it tends to push back due to its inability to makes sense of a situation. So, the bottom line is that an element of my brain, due to the inability to make sense of situations, has caused a lifetime of interesting results. No, I'm not a psychologist so none of this is intended for you to treat yourself. It is simply a personal life portrait painted of myself.

So, what makes a veteran resilient? In my case, the road back to life was in 1972 when Joyce and I were sitting in a church service and we both gave our lives over to God. During the years after, many aspects of our married life improved. Resilience can be defined as the ability to bounce back after setbacks. Some veterans appear to have found the secret and to others, it remains a mystery.

Although I have reconciled myself, I am still haunted by nightmares of my times in the field. I still suffer pain due to the complications of being shot down and crashing and being shot. Maybe even more so the memories of the time in the hospital. I am a very fortunate person in that I wasn't injured any worse than I was.

INTO THE STORM

I thank God, my wife, and family for every day that we're together.

13. Veterans Administration

AUGUST 1969 — 2018

Imagine 50 years of Abuse by an Organization with a Mission to Help Veterans

Synopsis of wartime injuries:

1. March 12, 1967, Helicopter UH-1D, 66-750, impact vertical crash—estimated fall 200± feet, Aircraft commander Cpt. Recher. All four crewmembers with complaints of back injuries.
2. Information on U.S. Army helicopter UH-1D tail number 65-12895

Date: 670401
This was a Combat incident. This helicopter was LOSS TO THEATER

This was a Logistics Support mission for Re-supply, to Forward Area. (No, it wasn't, it was to pay the soldiers in the field)

Unknown this helicopter was on Take-Off at 0075 feet and 025 knots. South Vietnam

According to the Major that went back to the helicopter and got me the left pedal and picked up several spent projectiles, he said, "The helicopter took over 80 hits"

The Army Goldbook said that the aircraft had 12 hits from: Small Arms/Automatic Weapons; Gun launched non-explosive ballistic projectiles less than 20 mm in size. (7.62MM)

The helicopter was hit in the Cockpit, fuselage, Cabin area, Engine compartment.

Systems damaged were: HYDRAULIC SYS, OIL SYS, ELECTRICAL SYS, TRANSMISSION, COMM SYS, ENGINE, PERSONNEL, FUEL SYS

Protection of PERSONNEL by Armor was Not Effective

Casualties: 02 WIA

The helicopter Crashed. Aircraft is later recovered by any means other than its own power.

Both mission and flight capability were terminated.

Crewmember had various injuries associated with this type of crash landing

The latter simply sets the stage of a scenario that any combat veteran may have encountered during one's tour in Vietnam.

Prior to being released from active duty, I applied for work with the Dothan Police Department, Houston County Sheriff's Department, and the Alabama State Troopers. I was denied employment on the physical status of the wounds received in Vietnam. Plenty of office jobs would be open, however, I was never "qualified" for one. Therefore, I began looking for anything. Soon to be unemployed I turned to the VA for help and assistance. At this point, I was unaware that the quasi-mission of the VA was to "Discourage-Delay-Deny" (D^3) veterans of any help without persistent annoying challenges presented by the VA personnel. As I began the process, I was sent to Atlanta, GA where it was determined that the disabling injury "Gunshot wound of the left foot, incurred during the Vietnam Era with a disabling degree of 20% effective 8-16-69.) It was then that they assumed that if they had specially made shoes for me that would solve my problem. Now granted it was tolerable so long as I didn't have to perform any prolonged standing, walking, bending over, crouching, running, climbing, and so forth. Or so that was my profile while on active duty. On my exit physical from the Army, three flight surgeons, standing in the break area drinking coffee, asked me to walk 10 steps in combat boots and based on that it was determined that I was ok.

After Atlanta, for a couple years, I was moved to Montgomery, Alabama, Perry Hill where I was informed that my medical records had been lost. That they were going to attempt to reconstruct them. Being that medical records were not stored in St. Louis, I didn't need to worry about the "fire" that occurred there. However, whenever requesting records from St. Louis was like pulling teeth to get the corroborating information required for benefits or placement within the VA system. Perry Hill, what a dump it was in the 1970's and 1980's, constantly misplacing records, unable to schedule appointments, and completely unknowing why the veteran was even at the appointment. I was in awe at how they could be so busy. Why are they so busy and with whom? It was known at that time that only 300,000 thousand had been wounded and only 153,303 had even required any hospital stay. I couldn't understand why appointments were an issue.

For years leading up to 1991, I had problems with metal, bone, and sores on the foot. Then in 1991, one of the Doctors decided I needed surgery on the foot and that it needed to be done now. They scheduled the surgery and then during the pre-op, they, after five attempts, couldn't get the compression tourniquet to work. Very painful indeed. They had to clean-up some of the broken bone, remove the middle toe and sew it back together. Rather simple surgery. They wouldn't let me stay in the hospital to recover since the VA would have to pay me on a much higher scale while I recovered. Joyce had to have the seats removed from the van, a recliner put into the van, and bring me back to Dothan from Montgomery. There she had to work a fulltime job as a 2nd grade teacher, tend to two children, and now a husband that was in bed recovering from surgery for six weeks. Seems fair, doesn't it?

After the surgery, I required an insert or prosthetic device between the opposing toes, so it would keep them from "migrating" to the middle and thus creating the same problem that the surgery was to correct. Where was I to get the prosthetic device? No one seemed to know or even care. Suggestions included putting gauze between them, rolling tape and placing it between the toes, cotton balls and then it was recommended that I "whittle" a small wooden block and sand it to fit. Now that set me off. This thing wouldn't be a problem if I just had an organization that was proactive and would practice the discipline of medicine.

So, I heard of something called a VA outpatient card in the 1990's. This would allow you to go to your family doctor or a specialist that could work with your disability. Sounded like an excellent plan to me. Well, it worked for a while and must have worked too well. For it was taken away and cloaked in secrecy for years. So, there <u>were and are</u> entitled groups within the VA system that <u>were and are</u> receiving preferential treatment. I couldn't understand why a combat wounded veteran couldn't receive specific treatment and medicines for combat related injuries.

Just after 2000, the shoes that I had been receiving from the VA for decades simply stopped. I received a letter from the VA that simply stated that shoes would no longer be provided. Then I received a tin that included every document they had and the shoe forms. That was it! No more specially made shoes for you to walk in. Now, what do you do? Call someone, so I call Montgomery VA and ask what am I to do? They don't know and can't help me.

Well if they can't help me, who can? So, they send me to a Podiatrist. He cautions me that if I make too much out of this that they may want to do surgery and cut off the front half of my foot. I ask for shoes and the VA says, "NO, BUT we can cut off part of your foot." What can I say 51 years later, déjà vu? I'm not having this conversation again and going through surgery at 72 years of age. The medical bus has left the building, you can prescribe pain medication from now on. Oh, that's right we can't have pain medicine because it's a level something. I have chronic pain resulting from crashing one helicopter and being shot down in another helicopter and wounded. What does it take? Please tell me.

I have learned to hate, detest and despise the VA. It's contractors, which are nothing more than medical mercenaries, condoned by the government and the VA employees. I am currently rated at over 150% of which only 80% by the VA voodoo math computations are compensable. After fifty years of being shuffled here and there, video-tele-therapy, doc-in-the-box failed concepts and annual visits resulting in no healthcare, withheld medications, screwed up pharmacy practices, and no continuity. If they aren't going to read the medical record, why collect the information? I've had five psychologists in just five years. I saw a Doctor this year (2018) for the first time in over 15 years. I've requested to see a "real" doctor every year or at least be referred to a specialist. I've requested referrals to Pain Management each time that I've been seen at my local clinic. I've had to reopen my claim eight times in an attempt to receive the help that should have been readily available only to have the VA mission statement of "Discourage-Delay-Deny" (D^3) thrown in my face. This is a caustic topic and thus I have maintained that the following be the solution for the VA system and more importantly the veterans.

Whereas, citizens of the United States and/or its territories choose to or are compelled to serve in the military in order to preserve and protect the constitution, and

Whereas, said participants that receive combat injuries should be entitled to receive a government identification card using RFID technology containing a military picture, identifying information and medical information upon discharge.

Whereas, the government identification card will serve as a medical payment card for the treatment and hospitalization of their healthcare and disability payments.

Whereas, disability percentage and computational awards will be based solely upon the individual's ability to perform his duties in the military MOS in the closest civilian occupational position. Therefore, if the individual is no longer able to perform aviator duties in civilian circles then disability is 100%. With no other attachments, stipends or accouterments.

Whereas, if the individual is required to wear, use, attach or describe in any other way apply prosthetic devices they will be included as healthcare related expenses.

In other words, limit who receives awards through fair practices, eliminate the VA healthcare system completely and standardize the benefits system. It will lower costs to the government and fairly serve veterans by recognizing those that need help due to their encounter with combat related injuries.

Do I consider PTSD a combat disabling condition? "CAMBRIDGE, ENGLAND—Soldiers have been suffering from post-traumatic stress disorder for at least 3,000 years, according to a paper written by Jamie Hacker Hughes, director of Anglia Ruskin University's Veterans and Family Institute, and psychiatrist Walid Abdul-Hamid of North Essex Partnership University NHS Foundation Trust. Historians often cite Herodotus' account of Epizelus, an Athenian spear carrier who experienced psychological problems after the Marathon Wars in 490 B.C., as the first recorded case of PTSD. But texts from Mesopotamia's Assyrian Dynasty (1300-609 B.C.) record traumas suffered by soldiers who were called upon to fight every third year during their military service. The symptoms were thought to have been caused by the spirits of the enemies whom the patient had killed in battle. "Ancient soldiers facing the risk of injury and death must have been just as terrified of hardened and sharpened swords, showers of sling-stones or iron-hardened tips of arrows and fire arrows. The risk of death and the witnessing of the death of fellow soldiers appears to have been a major source of psychological trauma," the paper reads. "Moreover, the chance of death from injuries, which can nowadays be surgically treated, must have been much greater in those days. All these factors contributed to post-traumatic or other psychiatric stress disorders resulting from the experience on the ancient battlefield." To read about the dramatic consequences of a battle in the Iron Age, see "The Price of Plunder."

Does it matter what I think? Seems to me the precedent was set and recognized for over 3,000 years.

https://www.archaeology.org/news/2922-150126-ancient-world-ptsd

Of the organizational disasters that plague the United States citizenry, the VA is the most shameful failure. No country should treat its veterans the way America has treated its veterans. We need to fix the problem of the Veterans Administration. Or at least change its name to something less disgraceful. I leave you with this thought... if the VA was doing its job would there be a need for any other outside organizations to be supporting veterans? DAV, VFW, American Legion, MOPH, Wounded Warrior Project, and so on. All these organizations demonstrate in one way or another the VA is a failed concept and has been since its establishment.

303,604 wounded—Purple Hearts issued during Vietnam to the living. Even if the VA needed to see all of them, how hard could it be to establish a program to do so. I would love an opportunity to help assist the VA in rectifying its tragic failure.

153,303 required a hospital stay

150,301 did not require a hospital stay

?? Non-combat\Accidental injuries needing and requiring a hospital stay

?? Sicknesses requiring a hospital stay

Accidents during training and daily work add up to the numerous personnel that requires long-term healthcare through the VA system.

For those veterans seeking benefits, consult with your local (VSO) Veterans Service Officer and if that doesn't work. Find out which veterans groups meet in your local area and are having success filling claims.

Closing Comments on Veterans Benefits

https://benefits.va.gov/benefits/

https://www.myhealth.va.gov/

My Closing comment, it didn't end with Vietnam, we are still "Ghosts and Dead Men Walking!"

References:

[1] http://en.wikipedia.org/wiki/David_H._McNerney

National Archives

https://www.archives.gov/research/military/vietnam-war/casualty-statistics

About the Author

A current picture of the old man

MR. DONALD RAWLINSON, BORN IN KALAMAZOO MICHIGAN IN 1946, IS THE SON of Francis and Joan Rawlinson. He was an all-round gymnast in High School and an avid electronics enthusiast.

Growing up all over Lower Michigan as a civilian dependent, Mr. Rawlinson turned his back on an offer (the draft) to be an infantry soldier and volunteered to join the Army. Upon being accepted in 1965, he joined the New Action Army for FUN, TRAVEL and ADVENTURE.

DONALD RAWLINSON

And, as a New Action Army Warrant Officer Aviator, he was given an all-expense-paid tour of sunny Vietnam in 1967. Mr. Rawlinson was wounded on April 1, 1967 and was medically evacuated in April to Japan. He received the Distinguished Flying Cross, Purple Heart and 4 awards of the Air Medal and the Vietnam Cross of Gallantry with Palm during the three months of service in Vietnam.

After four years of service, Mr. Rawlinson joined the civilian workforce and attended college. There Mr. Rawlinson graduated with an emphasis in Management and Computer Science. He worked in a manufacturing environment as a Purchasing Agent, Foreign Steel Buyer, Specialty and Exotic metals procurement, Materials Manager and Comptroller. As with many an Army Aviator, Mr. Rawlinson had a difficult time determining what he wanted to do when he grew up. He has held positions in the computer industry in design and development, working with Universal Energy Systems, Paravant Computers, Harris Corporation, and Wright State University on the design and development of a 1556 data bus analyzer for military vehicles. Mr. Rawlinson served as the consultant to the Doctoral staff, Wright State University, on the design of primitives for the 1556 analyzer. He has held associations with the Harding Think Tank, GM Microcomputer Division, Atlanta Metropolitan Area Rapid Transit Authority and San Francisco Bay Area Rapid Transit Authority on automation and rail maintenance technologies. He has worked on the Boeing Sikorsky design project for the LHX (Comanche) project, specifically the cockpit design.

In 1986, after teaching computer science courses for six years, Mr. Rawlinson was entreated to join the civil service workforce thinking that he could be a positive influence. This move led Mr. Rawlinson into a very exciting period of introducing technology and software engineering to the military community. His goal has been to deliver products and services to the customer that improve, sustain, and effectively support the soldier. He joined the United States Army School of Aviation Medicine (USASAM) in 1996 and is certified by the military as a Level

II Systems Administrator and Network Manager. Mr. Rawlinson has received four awards of the Commander's Award for Civilian Service, two awards of the Army Achievement Medal, is the recipient of the Superior Civilian Service Medal.

Mr. Rawlinson retired from the Department of Defense (USASAM) in April of 2007. His interests in retirement are his family, church, and home. Not wanting to be idle, he and Dr. Faulk have pursued the establishment of the Alabama Law Enforcement Alliance for Peer Support. A non-profit organization for training of First Responders in Peer Support.

He is married to the former Helen Joyce Dukes of Dothan, Alabama and has two daughters, Michelle (Mike) and Heather (Tim). Their love and devotion defy all logic. Both of them have had the patience of saints and have hung around out of sheer curiosity as to see what the village idiot is going to do next. Mr. Rawlinson has four grandsons (Jamie, Jeremy, Chris, and Isaiah) and one granddaughter (Sarah).

www.ingramcontent.com/pod-product-compliance
Lightning Source LLC
Chambersburg PA
CBHW030111100526
44591CB00009B/363